Advances in Intelligent Systems and Computing

Volume 946

The series "Advances in Intelligent Systems and Computing" contains publications on theory, applications, and design methods of Intelligent Systems and Intelligent Computing. Virtually all disciplines such as engineering, natural sciences, computer and information science, ICT, economics, business, e-commerce, environment, healthcare, life science are covered. The list of topics spans all the areas of modern intelligent systems and computing such as: computational intelligence, soft computing including neural networks, fuzzy systems, evolutionary computing and the fusion of these paradigms, social intelligence, ambient intelligence, computational neuroscience, artificial life, virtual worlds and society, cognitive science and systems, Perception and Vision, DNA and immune based systems, self-organizing and adaptive systems, e-Learning and teaching, human-centered and human-centric computing, recommender systems, intelligent control, robotics and mechatronics including human-machine teaming, knowledge-based paradigms, learning paradigms, machine ethics, intelligent data analysis, knowledge management, intelligent agents, intelligent decision making and support, intelligent network security, trust management, interactive entertainment, Web intelligence and multimedia.

The publications within "Advances in Intelligent Systems and Computing" are primarily proceedings of important conferences, symposia and congresses. They cover significant recent developments in the field, both of a foundational and applicable character. An important characteristic feature of the series is the short publication time and world-wide distribution. This permits a rapid and broad dissemination of research results.

**** Indexing: The books of this series are submitted to ISI Proceedings, EI-Compendex, DBLP, SCOPUS, Google Scholar and Springerlink ****

More information about this series at http://www.springer.com/series/11156

Michele Moro · Dimitris Alimisis · Luca Iocchi
Editors

Educational Robotics in the Context of the Maker Movement

Springer

Editors
Michele Moro
Department of Information Engineering
University of Padua
Padua, Italy

Dimitris Alimisis
European Lab for Educational Technology
Edumotiva
Sparti, Greece

Luca Iocchi
Department of Computer Control
and Management Engineering
University of Rome "La Sapienza"
Rome, Roma, Italy

ISSN 2194-5357 ISSN 2194-5365 (electronic)
Advances in Intelligent Systems and Computing
ISBN 978-3-030-18140-6 ISBN 978-3-030-18141-3 (eBook)
https://doi.org/10.1007/978-3-030-18141-3

This Springer imprint is published by the registered company Springer Nature Switzerland AG
The registered company address is: Gewerbestrasse 11, 6330 Cham, Switzerland

Preface

The volume "Educational Robotics in the Context of the Maker Movement" includes papers presented at the International Conference "Educational Robotics in the Makers Era-EDUROBOTICS 2018" held on October 11, 2018, in Rome. The conference was organized by the *Dipartimento di Ingegneria Informatica, Automatica e Gestionale "Antonio Ruberti"*, University *"La Sapienza"* of Rome (Italy, http://www.dis.uniroma1.it/en), EDUMOTIVA (Greece, www.edumotiva.eu), the *Dipartimento di Ingegneria dell'Informazione*, University of Padua (Italy, http://www.dei.unipd.it/en). The conference was the 2018 edition of a sequence of scientific international meetings of researchers working in the Educational Robotics (ER) field, starting from the 1st International Workshop entitled "Teaching Robotics and Teaching With Robotics-TRTWR" organized in Venice in the context of the TERECoP project (www.terecop.eu), followed by other TRTWR workshops (Darmstadt 2010, Riva Del Garda 2012, Padua 2014), and then generating the first edition of the "Educational Robotics in the Makers Era" conference held in Athens in 2016. Publications from those workshops have included open online proceedings, two special issues in the journals: Themes in Science and Technology Education, 2013 and Robotics and Autonomous Systems Journal, 2016, and the book "Educational Robotics in the Makers Era" (Springer 2017).

This book explores more deeply the connection between educational robotics and the maker movement which is currently inspiring new innovative ways to exploit technologies as educational "vehicles" for new ways of learning in and out of school, with a special focus on block-based programming environments, Do-It-Yourself electronics, 3D modelling and printing, the use of intelligent distributed systems, the IoT technology, gamification, and more. So empowered, Educational Robotics (in a broader sense) is proving a more and more fruitful field of development and experimentation both in formal and informal education settings, involving a continuously growing community of researchers and educators with interest in this field across EU and beyond. Starting from pedagogical premises that are the lessons learnt from constructivism, constructionism and social constructivism (Piaget, Papert, Vygotsky), current advances in Educational Robotics focus on the development of STEAM (Science, Technology, Engineering, Arts and

Math) education and the 21st-century skills: creativity, critical thinking, team working, and problem solving.

"Learning by doing" and "Learning by making" are the "mantras" around which the Maker Movement and its inspiring principles are successfully developed. Low-cost materials, free software, and a low access threshold are the pre-conditions to democratize access to new opportunities for learning through the 21st-century digital making technologies. The production of meaningful artefacts, where the "robotic" component can be more or less relevant, is an up-to-date interpretation of those "powerful objects to think with" that Papert considered crucial means for an effective and long-lasting education. The 2018 EDUROBOTICS conference was a valuable occasion to deepen these aspects with scientific contributions that deal with new proposals and innovative experimentations in this field.

The book includes 14 chapters plus six short papers coming from twelve different EU countries, the USA, and Israel.

The content of the book is organized into five sections

First section: Educational Robotics and the Maker Movement
Second section: Pedagogy behind Educational Robotics
Third section: Programming, Games, and Social Robots
Fourth section: Technologies for Educational Robotics
Fifth section: Short Papers Reporting Good Practices or Work in Progress (Presented in the Conference as Posters)

We thank the conference participants, academics, researchers, and educators from all the levels of education (primary, secondary, and tertiary), and the young researchers, PhD and postgraduate students, for their active participation and great contribution to the success of the conference and for authoring this book. Special thanks go to our Programme Committee members who have reviewed the papers and provided important help to authors to improve their manuscripts.

Finally, this book is dedicated to the memory of Edith Ackermann, the brilliant Swiss-American psychologist, collaborator of Seymour Papert at MIT and pioneer in studying the influence of technology on children's learning, who passed away at the end of 2016.

March 2019

Michele Moro
Dimitris Alimisis
Luca Iocchi

Organization

Programme Committee

Dimitris Alimisis	EDUMOTIVA (European Lab for Educational Technology) Greece
Richard Balogh	Slovak University of Technology in Bratislava
Ansgar Bredenfeld	Dr. Bredenfeld UG
Dave Catlin	Valiant Technology
Baran Cürüklü	Mälardalen University
G. Barbara Demo	Dipartimento Informatica, Universita Torino
Amy Eguchi	Bloomfield College
Nikleia Eteokleous	Frederick University Cyprus
Alessandro Farinelli	Computer Science Department, Verona University
Paolo Fiorini	UNIVR
Luca Iocchi	Sapienza University of Rome
Ilkka Jormanainen	University of Joensuu
Ken Kahn	University of Oxford
Chronis Kynigos	Educational Technology Lab, University of Athens
Lara Lammer	ACIN-Vienna University of Technology
Wilfried Lepuschitz	Practical Robotics Institute Austria
Monica Malvezzi	University of Siena
Emanuele Menegatti	University of Padua
Emanuele Micheli	Scuola di Robotica Genoa Italy
Stefano Monfalcon	Town Museum of Rovereto
Michele Moro	University of Padova
David Obdržálek	Charles University
Pavel Petrovic	Comenius University
Alfredo Pina	Public University of Navarra

A. Fernando Ribeiro	University of Minho
Sandra Schön	Salzburg Research Forschungsgesellschaft
Gerald Steinbauer	Graz University of Technology
Calkin Suero Montero	University of Eastern Finland
Jeffrey Too Chuan Tan	The University of Tokyo
Igor Verner	Technion.ac.il
Markus Vincze	Vienna University of Technology

Contents

Educational Robotics and the Maker Movement

Designing a Makerspace for Children – Let's Do It

Sandra Schön[1], Martin Ebner[2(✉)], and Maria Grandl[2]

[1] Salzburg Research, Innovation Lab, Salzburg, Austria
[2] Educational Technology, Graz University of Technology, Graz, Austria
martin.ebner@tugraz.at

Abstract. When makerspaces are designed for children, special motivation and reasoning needs to be made. Within this article, we describe a case study: A temporary four-day open makerspace for about 40 children per day. Motivation, considerations and the development process as well as the actual realization are described and discussed. We comment on how such a space for children and adolescents should be arranged for future studies. As described, considerations in terms of participation, peer tutoring and gender mainstreaming influence the design of the space, the methods used, as well as the general setting.

Keywords: Maker education · Maker Days · Maker movement · Children · Education

1 Introduction

Makerspaces, fablabs and hackerspaces are typically designed with adults in mind. Special offers are made for children occasionally, but adults are usually seen as the target group and considered to be the main users. In recent years, making activities for children have gained more interest. Seen as an activity with "creative and technological impacts, learning by making is an important component of problem-solving and relating educational content to the real world" [1]. Making is supported by its own learning theory: Seymour Papert proposed "constructionism" or "learning-by-making" [2] where learners would use tools to make things to construct knowledge. Papert, who died in July 2016, left his marks in several popular maker tools for children such as Logo, Lego Mindstorms, Scratch and the MaKeyMaKey and, additionally, in books such as "Invent to Learn: Making, Tinkering, and Engineering the Classroom" by Martinez & Stager (2013).

Although interest in making with children has increased, makerspaces for children are not common in the whole world. In the USA, makerspaces for kids exist in various learning environments: in-school, after-school, home-based and museum-based. [3] The MENTOR program aims to have more than one-thousand makerspaces installed in high schools [4]. In Europe, makerspaces with children as a single or main target group are less common and less studied.

Within this contribution, a case study of special makerspaces for children is described and discussed. The temporary, four-day long open workshop for about 40

M. Moro et al. (Eds.): Edurobotics 2018, AISC 946, pp. 3–15, 2020.
https://doi.org/10.1007/978-3-030-18141-3_1

children per day, called "Maker Days for Kids", was probably the first event in a German speaking country with an open approach and more than 130 visitors.

2 Methodology

To get an idea of how a proper maker space setting for more than 40 children and adolescents, aged 10 to 14, should be organized, including tools, activities, rules and design, we did a literature study. We looked at evaluations that were described in existing publications [5]. We combined these results with our own motivation and as responsible initiators and managers of the "Maker Days" project, we wrote proposals for financial support. In this paper, we will describe the motivation behind the project as well as the different stages of development that lead to our final setting and concept.

During the Maker Days, we used printed workshop cards to document how many workshops were offered and to monitor which participants attended which workshops. Every participant got an ID, that was used for all further documentation.

In this paper, we will present the results of the case study and reflect on the planning and realization of the Maker Days to provide the lessons learned for future events with a similar goal and target group.

3 A Temporary Workshop: The Maker Days for Kids

Considerations, steps of development, preparations and results of a four-day long maker space designed for children between 10 and 14 years old are discussed. The event "Maker Days for Kids" was created by Sandra Schön and supported by Martin Ebner.

3.1 Motivation

The event called "Maker Days for Kids" took place in April 2015 in a town in Southern Germany. In creating this makerspace, Sandra's idea was not only driven by the desire to provide an innovative workshop on the creative usage of new and innovative digital tools. Taking her background as a researcher into account, her goal was to investigate how such a setting could be organized and to answer the following question: Does this setting attract children of both sexes and from families of less fortunate social backgrounds?

3.2 Early Considerations

The first time we got in touch with the maker movement was in a German speaking open online course called "mlab14: Kinderzimmer Production 1". The aim of the course was to show children how to use technologies for learning purposes and creative work. Two girls, with whom we had worked together in former projects, also contributed to this course with a presentation about two projects in which they were involved. They talked about video production with tablets (and how they brought the mayor to Paris with the help of the green screen technology) and described how they produced a stop-motion film about monsters that appeared in the city's center.

Another part of the course was about the "Maker Movement" and we were absolutely stunned. Everything that was mentioned there was exactly what our work with children was already about: Doing something creative with the help of new technologies, even this did not include 3D printing or laser cutting at that time.

As researchers, the next step was to collect materials and literature and to write some posts on a German speaking weblog about media education and maker education. Together with Swapna Kumar, we also published a longer scientific contribution in the European e-learning papers [1].

While discussing the idea of a bigger making event with children and adolescents in an open setting, a friend suggested a possible call for projects. While waiting for a decision, we had already started a weblog for the project and announced that we would try to organize the event called "Maker Days for Kids" in autumn 2013 and asked for sponsors and additional co-operations. When we received the rejection for the project funding, we openly shared the decision with a post in our personal weblog as well as in the newly created project's weblog.

The reaction to these blog-posts was fantastic: A potential sponsor, the HIT-Stiftung, called (there was no contact before). They wanted to learn more about the idea and asked for the minimum amount of money needed to run the makerspace.

With the help of this funding, which easily covered all costs for material and rent but didn't cover the effort involved (especially the countless hours in preparation) the first step was done.

The next step was to find a proper location – the local youth center was very interested and agreed, that the event could be held there. Martin, who also works as an educator in educational technologies, organized part of his team from Graz University of Technology and some students to support the event. Seen as a unique learning event, they were all excited to be part of it. Sandra also asked her boss for support and other partners were found as well.

Nevertheless, the development of the makerspace, the design of the setting as well as the planning and preparation of training events, had still to be done.

3.3 Initial Design Considerations

As researchers, we perused examples and descriptions [3, 4, 6] in the first stage of the development process. Nevertheless, we could not find helpful descriptions for an open setting, where children could come and go without registering or applying beforehand. Announced as a requirement for funding the project, one of our main sponsors, the HIT-Kinder-Stiftung, made it clear that the Maker Days should address children from less fortunate social-economic backgrounds.

Additionally, it was hard to decide on the tools we should use and, on the projects/activities we should provide. Finally, we had to develop a setting and an evaluation design which allowed the documentation of the activities within the makerspace.

As a first step, we wrote down six principles that we wanted to follow within the design and realization stages: We wanted (a) to realize an open and low-threshold offer

(b) to foster participation of children, (c) to support idea and innovation development, (d) a (self-organized) competence development concerning new media and ICT (e) to design a gender sensitive setting (gender mainstreaming) and (f) to select tools and software that children could continue to use after the four days.

The development of the makerspace was the most important and longest part of the project. All in all, it took us six months. Finally, we shared and discussed our ideas with a group of experts (called Maker Days experts' committee), teenagers (peer tutors) and we organized a two-day long introduction meeting for all peoples involved. There we openly discussed minor issues and applied smaller changes. Additionally, the project consisted of a rising number of organizations collaboratively working together, e.g. the local youth center, two research institutions, the regional young researcher center (in German "Schülerforschungszentrum"), as well as companies and individuals who provided materials and/or support.

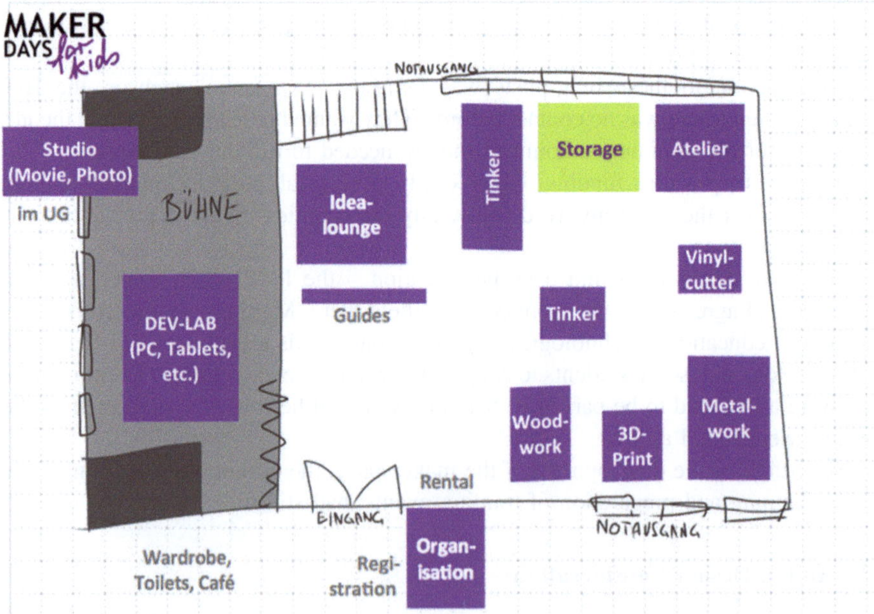

Fig. 1. Room drawing of the Makerspace, where the "Maker Days for Kids" took place

3.4 Designing the Makerspace Location

Although we were developing an open setting, the organization and preparation was complex (see Fig. 1).

Design Evolvement. From the beginning, the local youth center made the main hall available for our project, including the use of the toilets and the main kitchen and some additional space like the room under the stage.

Three peer tutors, Sandra and another tutor visited the location. First considerations were:

- Where are plugs? Especially: Where is more than one plug and is it possible to connect multiple sockets?
- Where should we install the router ("Freifunk", free Wi-Fi) so that we can easily connect to the main router in the youth center's office?

This technology-driven approach led to the conclusion that all the equipment that needs a lot of electricity should be placed on the stage. At the first moment, this sounded a little bit weird because we only considered the main hall as a location for the Maker Days.

During this first visit, Sandra realized that the whole hall had to be made more attractive and engaging, with separated spaces for certain activities etc.

The next design questions were:

- What are low-cost possibilities to mark certain spaces? (for example buntings, big sign posts, tables, shelves if available, fabrics)
- What furniture and equipment do we need? (How many tables and chairs? Where should we place them? How many computers?)
- What do we have? (e.g. the youth center has a sofa)

Building upon the parallel ideas on equipment and activities (for example 3D printing, making videos, using colors, wood work and need of simple wood tools) a sketch of the room was done. This sketch did not differ a lot from the room drawing in Fig. 1.

Especially the concept of the "idea lounge" as heart of the maker space made it necessary to develop a setting which is interesting for children and teenagers. Besides the existing sofa, we therefore bought several beanbags.

As far as possible, all decorating as well as functional furniture and equipment was bought or painted in the project's colors (light green and violet).

Decisions about the room and its equipment were also discussed in the preparation sessions with the peer tutors, which also helped in painting posters, signs, etc.

Looking back, one of the main learnings was that it was useful to start with the main technical challenges as this produced an unusual but smart solution with the "developers' space" on the stage. We should have thought about the possibility that a lot of stuff, especially paintings, need a place to dry, that it would have been useful to have a sink in the room for cleaning hands and that children need "save places" for their products as we experienced that someone missed a 3D printed object or a LED construction.

Room Plan and Tools. The Maker Days for Kids was offered in the Youth center of a small city (less than 20.000 inhabitants) in southern Germany, near the Austrian border. The big hall of the center with approximately 300 m^2, the stage, a little accompanied room and the cellar under the stage were used. Additionally, the participants had access to the kitchen (for example to prepare their lunches) and additional rooms.

The big hall was completely re-organized as different areas. Therefore, tables, shelves, posters with the area's name and decorative stuff as buntings were used to organize the areas.

At the registration desk, everyone had to check-in. There, the participants could also ask for support in organizational issues and for help if something got lost or if they needed special tools. At the center of the main hall, there was an information desk displaying the daily events (e.g. visits of the press, the daily challenge) and the mini-workshops. Everybody was invited to offer or participate in these mini-workshops. They usually lasted no longer than 20 min and were intended as a first introduction for the use of tools, software or techniques. The big table also acted as meeting point for the Maker Days tour for newcomers (and sometimes adults, especially parents), that were guided by tutors and peer tutors.

The other spaces in the hall included: (a) a space for woodwork with a set of basic tools, (b) two 3D printers, (c) a place for electronic tinkering ("metal work"), including work with LED and a storage place for material (with no direct access by the participants), (d) a table with the vinyl cutter and the ironing press, (e) the atelier with a sewing machine, sewing stuff and tools and a big basket of used clothes, (f) a huge supply bank, freely available to each participant, including paper, canvas, colors, cutters, glue, (g) two big tables for tinkering of all kinds (one especially for dirty work), (h) the "idea lounge", a comfortable space with a couch, bean bags, books and white boards (i) the "dev lab", situated on the stage with about 9 laptops and 3 tablets, the Lego Mindstorms robots and a drone. Under the stage, there was a helpful additional room: The darkened room was used for the light painting technique and video production.

3.5 Designing the Makerspace Activities

Design Evolvement. Besides Sandra and Martin, the peer tutors were also involved in discussing which activities would be interesting and how they can be arranged so that newcomers would easily become active users of the makerspace.

The design questions were:

- What activities are interesting?
- How could users get easily involved with the makerspace's different sections as well as activities? (we developed the idea to start with a guided tour for every new beginner and a "waiting space" where the new participants were encouraged to start tinkering with their badge and a small paper journal for notes)
- What are our rules? How would we like to work? (Therefore, we adapted the maker manifesto [7]).
- How can we prevent the usage of very expensive materials or limited tools? (Some materials were limited. Participants had to ask for some materials and tools at the registration desk. Only one 3D-print was possible every second day, for example.)

Building upon the idea of creating an e-portfolio for every participant, Sandra also came up with the idea of badges. Every badge was attached to a lanyard and contained the participant's ID. These ID's were also used to tag photos and posts within the weblog, so that everyone had the possibility to share his/her own e-portfolio (The ID was also mentioned in their certificate).

The development of the design was very tricky as we did not find any good examples of similar open makerspaces for children.

Invitation. Conceptualized as an open setting, the participants did not have to register for the event and participation was for free, of course. Concerning gender issues, we developed posters and materials, which did not highlight technological issues such as programming, 3D-printing or video production. From the literature study and our own experience, we knew that this kind of announcement usually attracts more boys than girls. We decided to keep it as inclusive as possible by just announcing "creative production within an open workshop".

The Maker Days' Principles. Together with some children, we provided some "Maker Days Principles" that we printed out several times and introduced to every newcomer. The principles are a variation of the Maker Manifesto by Hatch (2013) [7]: "1. Be creative! Make it different; try it! 2. Be curious! What are the others doing? 3. Copy! Good ideas are meant for copying. 4. Share! Share your ideas, materials and tools! 5. Work sustainably! Use garbage, avoid waste! 6. Ask for help! Look for support! Ask! 7. Fail! It does not hurt. 8. Tidy up! (A stupid principle has to be there, too) 9. Have fun! And now start!" Another important rule was not mentioned in the list of principles but was put above the entrance: "No adults allowed!" One exception was the special "adult's tour", run by peers or adults, where adults were asked to stay with the tour guide.

The Maker Days' Offers. Designed as a participative event, the main focus was on the participants' ability to self-organize their activities in the maker space according to their interests (see Fig. 2). In general, participants could use all materials and tools (if available). Free WiFi, special printed-out learning materials (e.g. for the programming language Scratch) as well as books were available for inspiration and tutorials. Within the DevLab, continued support from tutors was available. For the vinyl cutter and the 3D-printer, there were regular time slots for printing (more or less organized by the tutors). Photos of the final results or projects were done at the registration desk and were included in the project's weblog.

Besides this, every newcomer got a guided tour through the whole makerspace, including an introduction about the organization of the day and the Maker Days' Principles.

Every day we did a "daily challenge". This was a special task, announced and moderated by the tutors. On the first day, the name of the event "Maker Days" was formed with cardboard letters and the participant's task was to design these letters. Later, the cardboard letters were used as piano keys together with the MaKeyMaKey kit. The challenge for the second day was to make a film about the Maker Days. On the third day it was a "Maker Days Game" and on the fourth day, participants were asked to help with the preparation for the final public presentation.

Mini-workshops were offered almost over the whole day, excluding the "challenge time" and the last half hour, which was reserved for tidying up. 120 mini-workshops were offered by (peer) tutors and more than 90 of these workshops actually took place.

Finally, every day ended with a common tidy-up-session and a presentation of the projects in the (overcrowded) idea lounge where the participants shared their experiences and results with the other participants and tutors. Opening at 10.30, the hall was closed at 16.30 every day.

Fig. 2. A part of the makerspace, looking from the "DevLab" into the idea lounge (left in front) and several spaces for material storage, tools and work (behind it).

Being a Participant in the Maker Days. From the perspective of a participant, the journey started at the registration desk, where some sheets of papers had to be filled out by the parents. Afterwards, newcomers got the materials to tinker their name badges and a small paper notebook. They were asked to wait for a peer tutor, who would do the guided tour. The waiting or tinkering time normally took some minutes. The peer tutors typically started the tour, when there were 3–5 newcomers. During the tour, the tutors explained every part of the space, including the rules, the possibility to eat, the location of the restrooms etc. Newcomers were also asked to start with one of the mini-workshops, which were announced at a designated pinboard, or with the daily challenge (always at 11.00 am without any parallel workshops). The most frequently asked question was whether it was all right to take stuff such as cutter knives and paint self-dependent without asking anyone. At the computer lab and the 3D-printer desk, tutors were available all the time to support participants and to invite potentially interested children.

As described in the next section, the children usually stayed for several hours and came back for another day (if not the next day).

4 Results and Reflections

4.1 Children's Participation and Activities

Usually, it is a complex task to document activities within an open setting in a big room and with such a huge number of people involved. Some developments and considerations made it easier to observe the main activities. First of all, we collected some general information about the children at the registration desk (e.g. age, sex, school, home town). Then, every participant got a name badge with an ID. The first number indicated the day of the visit. This was helpful to identify newcomers or peer tutors, which, for example, started with "0". The next two digits were the numbers they got at the registration. All three digits together resulted in a unique number and the ID was used for registration, the documentation of the mini-workshops as well as the name of a file (e.g. for printing or cutting at the machines). The ID was also used to keep track when limited materials such as the set of lenses for the virtual reality cardboard goggles were used. When participants borrowed special tools, we also recorded their ID. Of course, all comings and goings were recognized at the registration desk. Last, but not least, the photos of the projects in the weblog were tagged with the IDs of all participants involved. This way, everybody got an individualized but anonymous online portfolio.

This structure made it easier to document and monitor at least parts of the participants' activities, at least in the organized mini-workshops. The data was collected and analyzed (see Table 1).

In general, 61 children participated in the Maker Days, 10 more in the general projects, and 8 were actively involved as peer tutors on each of the four days. At least 8 adults (researchers, experts, teachers and teacher training students) served as tutors over the four days. All in all, 124 "daily visits" were counted (as several children came for more than one day). For example, a special 3D printing expert or an urban gardening expert joined for a single day.

Table 1. Activities of the participants (without peer tutors) in mean or in percent (N = 61)

Activities of participants	In figures
Percentage of children who participated in the Maker days only one day	39%
Percentage of children who participated in the Maker days for 2 days	25%
Percentage of children who participated in the Maker days for 3 days	13%
Percentage of children who participated in the Maker days for 4 days	23%
Number of visited mini workshops (average)	5,3
Percentage of children who participated in a mini workshop	90%
Percentage of children who constructed 3D models (for 3D printing)	39%
Percentage of children who used sheets for the vinyl cutter	43%
Percentage of children who borrowed the 3D doodler (others could have used it, too)	7%
Percentage of children who took lenses for the virtual reality cardboard goggles	11%

Of course, one of our main interests was to see how many children had participated. On the first day, 28 children took part, and our maximum limit of 40 was nearly reached on day 2 and 3 (39). On day 4, the Maker Days attracted 28 kids and teenagers.

The majority of visitors came for the opening and stayed until the end of the day. 15 children who had come on day 1 came again on day 2 – which is an impressive ratio of 86%.

The analysis also gives some insights into the activities. All participated in the guided tour and 90% of the children took part in at least one mini-workshop. On average, each participant took part in more than five mini-workshops. More than a third of all participants printed at least one 3D model and cut a shape of vinyl film with the vinyl cutter. The number of children who borrowed the 3D doodler was smaller (7%). 11% of the children used the lenses for their own VR cardboard goggles.

With the help of the registration data, additional results are available. Because we asked for the school, we can analyze the data for different types of schools. In general, this project attracted a surprisingly large number of students who came from disadvantaged schools (40%). 28% of all participants were from advantaged schools. The other participants attended an elementary school. If we assume that the choice of school provides information about the socio-economic status of the families, we potentially achieved our goals concerning the socially disadvantaged. The deeper analysis does not show significant differences between the analyzed activities regarding the three types of schools.

Additionally, we also tried to attract as many girls as possible, as they are usually not attracted by technical projects in large numbers. Therefore, the ratio of 44% of female participants is really satisfying. Looking at the difference between boys and girls we only identified one or two statistically significant differences: Only 21% of the girls but 53% of the boys developed at least one 3D model ($p < 0,01$). Only girls (14%) borrowed the 3D doodler ($p < 0,05$). We did not recognize other significant differences, also in other areas (for example on how the participants got into contact with the Maker Days, see [5]).

Looking back at the considerations and principles we are very satisfied with the results we have achieved during the Maker Days. It seems that we succeeded in designing a low-threshold offer and attracting nearly as many girls as boys, which was also true for our peer tutors. On the other hand, we saw only limited initiatives from the other children. Only one external (non-peer tutor) male participant offered a workshop on day four, for example. We think that this would have improved if the event lasted longer.

4.2 Programming Languages and Educational Robots

For coding activities, the visual programming language Scratch and the mobile app Pocket Code were used. Pocket Code is a free mobile app, which was developed within the non-profit Catrobat project at the Graz University of Technology. The aim of the app and the Catrobat project is to raise interest in coding and STEAM-related subjects and to acquire coding skills in an entertaining and engaging way. The app acts as a development environment for programs that can be created directly on the smartphone or tablet with access to all built-in sensors. The values of the inclination sensor, for example, can be used to change the position of the object on the screen. With the help of the loudness sensor, users can control the size or the rotation of an object with their voice. The possibilities are endless. The used block-based programming language was developed according to the model of Scratch and the commands have been translated in many different languages [8] Users can quickly write programs with the help of command bricks, which are categorized in command classes (e.g. control, event, sound, data or motion) and which differ in colour. Currently, interfaces are being created to control different robots or boards like the BBC micro:bit with the app. Thus, the app is getting even more powerful.

During the Maker Days, participants created their own games with Scratch and Pocket Code. For programming with Scratch, we printed out different "Scratch-Cards". These are short coding tutorials helping beginners to perform certain tasks. But only a few participants used these cards. We observed that many participants, who came to the DevLab for coding activities, preferred to ask their tutors for help instead of trying out different commands independently. In comparison to other workshops, the mini-workshops about programming with Scratch and Pocket Code did not attract a lot of participants. A reason for this could be the fact, that the participants could not orientate themselves towards a final product or result, like an 3D-printed object. Compared to other mini-workshops, the participants needed more time to get used to the development environments and to experience a sense of achievement. They came up with a lot of great ideas for their games, which were largely inspired by well-known computer games. But they soon found out, that it is quite challenging to turn their ideas into code.

Besides Scratch and Pocket Code, kids could participate in workshops with the Lego Mindstorms robots. All in all, only 4 girls participated in Lego Mindstorms workshops that were explicitly announced for girls. Only a few participants worked with the Lego Mindstorms robots as the assembling of the Lego bricks, according to the given instructions, is a time-consuming activity. As well as that, working with Lego Mindstorms robots requires explanations of the development environment and information on how to control the sensors. To sum up, for didactical settings like the mini-workshops within the Maker Days for Kids, we suggest using other educational robots, that require only a short introduction.

4.3 Design Failures and Issues that Led to Changes

We did not recognize anything like a complete design failure. We saw the Maker days as a pioneer project and we also reflected on and redesigned things within the development process. But of course, some ideas did not work as planned:

- The children's participation as guides for the newcomer tour was okay, but only one announced his/her own workshop (all other active children were peer tutors).
- The children's projects were not very ambitious or creative. We guess that this is a result of the broad range of activities and the fact that the prior knowledge was in general limited. We did not have "junior nerds" as participants.
- The idea lounge was not used as a space for getting new ideas besides the organized activities.

5 Conclusion

Some of the visitors of the Maker Days asked, if others might use these models in the future. We see possibilities but we also see a lot of restrictions.

Looking at the Maker Days, the events was planned, run and financed as a model project. As a regular event, a range of sponsors and supporters is needed to deliver a similar offer regarding tools, tutors and concept.

Of course, existing structures as maker garages or fablabs with opening hours for children or collaboration models with schools might be a loophole for successful implementation – but likely with a smaller impact.

Nevertheless, we see also a chance for the future. First of all, we are redoing the Maker Days in 2018 at Graz University of Technology aiming to foster STEAM-education as well as to get children involved in making activities. Our findings serve as a basis for other (smaller and bigger) projects, like the Horizon 2020 project "DOIT-Entrepreneurial skills for young social innovators in an open digital world" (2017–2020), co-financed by the European Commission. It builds upon the consideration that social innovations in makerspace settings allow authentic learning experiences fostering future entrepreneurial spirit and ambition to co-create a (better) world (see http://DOIT-Europe.net). Furthermore, we are thinking about how we can bring this form of education to schools. Therefore, a lecture called Maker Education was included in our teacher education curricula for handicraft teachers. In a long run, we have the idea to bring it to a formal school setting.

If someone sees the sketched makerspace settings as attractive for his/her work and wants to hear our most valuable lessons learned, we would give the following suggestions: Do it like a maker: Plan and make it yourself – and of course: with children [9].

Acknowledgements. We thank the HIT-Kinder-Stiftung for financial support of the Maker Days for Kids as well as all other supporters for their endless help.

References

1. Schön, S., Ebner, M., Kumar, S.: The maker movement. implications of new digital gadgets, fabrication tools and spaces for creative learning and teaching. In: eLearning Papers, vol. 39, pp. 14–25 (2014)
2. Papert, S., Harel, I.: Preface, situating constructionism. In: Harel, I., Papert, S. (eds.) Constructionism, Research Reports and Essays, 1985–1990, p. 1, Norwood (1991)
3. Young Maker: Maker Club Playbook (2012). https://makered.org/wp-content/uploads/2016/10/Maker-Club-Playbook_Young-Makers-Jan-2012-6_small.pdf. Accessed 30 Nov 2018
4. Makerspace/Maker Media. The Makerspace Playbook. School Edition (2013). http://makered.org/wp-content/uploads/2014/09/Makerspace-Playbook-Feb-2013.pdf. Accessed 30 Nov 2018
5. Schön, S., Ebner, M., Reip, I.: Kreative digitale Arbeit mit Kindern in einer viertägigen offenen Werkstatt. Medienimpulse, vol. 1 (2016). https://www.medienimpulse.at/articles/view/829. Accessed 15 Jul 2018

6. New York Hall of Science: A blueprint: Maker programs for youth (2013). https://nysci.org/wp-content/uploads/nysci_maker_blueprint.pdf. Accessed 30 Nov 2018
7. Hatch, M.: The Maker Movement Manifesto: Rules for Innovation in the New World of Crafters, Hackers, and Tinkerers. Mcgraw-Hill (2013)
8. Slany, W.: Pocket code: a scratch-like integrated development environment for your phone. In: The Companion Publication of the 2014 ACM SIGPLAN Conference (2014)
9. Panke, S.: What does it take to make a maker? an interview with Sandra Schön. In: Association of the Advancement of Computing in Education, 26 April 2016. http://www.aace.org/review/make-a-maker/. Accessed 30 Nov 2018

Enactive Robot Assisted Didactics (ERAD): The Role of the Maker Movement

Hagen Lehmann(✉) and Pier Giuseppe Rossi

Università degli Studi di Macerata, Macerata, Italy
{hagen.lehmann, pg.rossi}@unimc.it

Abstract. The aim of the presented work is to outline a theoretical approach to the integration of robots into didactic contexts, with a specific focus on enactive didactic processes. In the first part of the article we will discuss, respectively, the framework of social robotics and that of enaction. In the second section we will theorize about ways to combine these two frameworks into an effective "Enactive Robot Assisted Didactics". The final discussion will reflect on the central position that robotic design plays in this cross-fertilization, and the outstanding role that the maker movement can play in this undertaking.

Keywords: Enactive didactics · Social robotics · Human-robot interaction · Maker movement

1 Introduction

In the last decades the ideas of enactivism [1], situated cognition and embodied cognition have foreshadowed a significant transformation, which is now under way, in our understanding of what teaching is, and how the teaching process of the future could look like [2].

Heinz von Foerster, a pioneer of radical constructivism, and a precursor of enactivism, used to express the key idea of this perspective change by emphasizing the difference between two ways of asking questions during the teaching process [3]. The first involves questions for which the teacher has already an answer – "illegitimate question". The second involves "legitimate" questions, whose answers has to be searched through a cooperation between teacher and student(s), which engages them in a process that creates knowledge, rather than making them passively consider knowledge that was produced by someone else [4]. Illegitimate answers are non-responses because the question does not truly exist. They are not directed to solve the doubts of the students, but to address issues that previous generations have posed. The answers are for the teacher who already "knows" the subject to be taught, but not for the student. In this sense they enhance the teacher's authority as a member of a scientific community, but not her role as a part of the classroom system.

One of the corner stones of the enactive approach in didactics relies in the idea that knowledge is constructed during a teaching process, which dynamically changes all agents involved and the context in which it is generated. The enactive approach sees the learning situation as a complex, multidimensional system in which every interaction

M. Moro et al. (Eds.): Edurobotics 2018, AISC 946, pp. 16–26, 2020.
https://doi.org/10.1007/978-3-030-18141-3_2

between its different parts changes not only the dynamics, but also the parts that are involved. Translated into a teaching situation, this means that teacher and student change each other, as well as the cultural and situational context, which in turn changes them, during the process of learning. The mere fact that both teacher and student are physically present in the teaching situation constantly changes the situation, and, with it, the context of the knowledge generation. All the individual embodied specificities of teacher and student create a continuously changing environment in which the only constant are the interaction mechanisms between them. This idea stems not only from Varela's enaction, but also in its roots in autopoietic cognitive biology [5] and its long genealogy, which reaches as far back as the 1940s, to the beginnings of cybernetics and research on complex systems [e.g. 6, 7].

This theoretical legacy, with its emphasis on feedback, makes enactive didactics a suitable candidate for the integration with an area of contemporary robotic research that has been labeled "social robotics" [8]. This field is in fact one of the most prolific applied scientific fields using ideas from research strands of constructivism and embodied cognition, to which enaction belongs.

The goal of social robotics can be described as the development of mixed human-robot ecologies [9] in which robots function as situated embodied agents, achieving structural coupling with humans via appropriate social interaction behaviors. One if the most intriguing, yet hotly debated areas in this domain is "artificial empathy" [10]. The advocates of this approach propose to tap into the human disposition to anthropomorphize – to ascribe feelings, intentions and emotions to anything nonhuman [11] – in order to create, in humans, the disposition to socially interact with robots. According to this approach, this emotional connection could be used to stimulate changes in human behaviors – improve for example their social skills, motor skills and memory. Robots with first attempts to exhibit behaviors like this are used at the moment in elderly care [12], robot assisted therapy (RAT) for children with Autism Spectrum Disorder (ASD) [13], and in care homes with dementia patients [14].

The first part of this article will give a brief overview of what enactive didactics is, and discuss briefly the underlying concepts of enaction, structural coupling, and situated cognition. In the second part we will introduce the basic ideas and principles of social robotics, and give examples of their use in different applied fields. In the third part we will give suggestions of how to combine the two approaches into an "Enactive Robot Assisted Didactics". The last part will elaborate on the central position robotic that design plays in the process, and the outstanding role the maker movement can play in these undertaking.

2 Enactive Didactics

One of the key points of the enactive approach, as it has been illustrated [1], is that it discards the old distinction between internal and external factors influencing the development of a system that exists in an environment. It replaces this distinction with the concept of structural coupling between the system and its environment.

This structural coupling is defined as a continuous mutual process that co-transforms the patterns of activity of the system and its environment. Through the expression of reactions, intended to maintain or re-establish internal equilibriums, the individual changes its environment and the process it interacts with it. In this way the system and the environment need to be thought of as two aspects of the same process.

If we see humans as such systems, this concept also remains true for social interactions between humans. In this case the means of communication might be more direct – verbal and nonverbal – but the mere occurrence of their situatedness in the social interaction means that they change each other, and based on this dynamically ongoing changes also change the characteristic or context of the interaction. Transposed to didactics, it becomes apparent that the teacher influences with her presence the context of the teaching situation, the subject to be taught, and the student. The student in turn changes the teacher, and the context; and the context changes the teacher and the student. This procedural view of mutual influence shifts the focus of teaching away from the "What" to the "How", from the static content to the dynamic process (Fig. 1. – adapted from [15]).

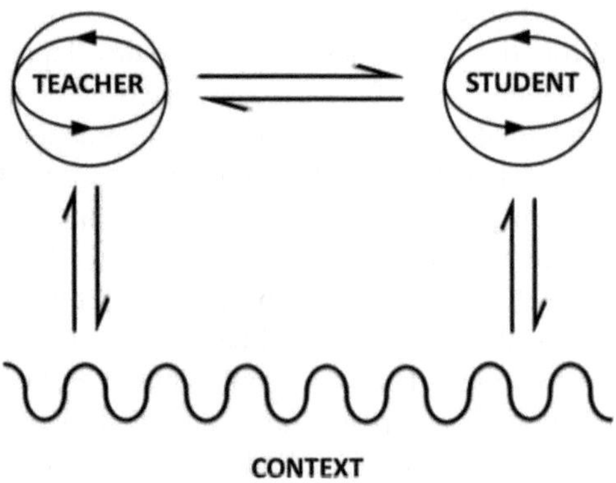

Fig. 1. Structural coupling between teacher, student and context in enactive didactics

From this perspective knowledge becomes a product of the teaching activity, the context of the teaching activity, and the individual characteristics of the teacher and the students, and the culture in which it is developed and used [16]. By following this line of reasoning it becomes clear, that it is no longer possible to distinguish between the content that is be taught, the way this content is taught, and by whom and to whom it is taught.

The enactive didactics approach focuses on the interaction between teacher and student in the process of creating knowledge. The central role of the teacher in this approach is to raise the awareness of a problem in the students. Having done this, it is possible in a next step to build together an answer. The way of constructing it is pre-outlined by the teacher as a mediator between the world of the student and the knowledge to be transmitted [17], and has the task of activating a cognitive conflict [18] that bridges the student's knowledge, the new problems to address and related new knowledge.

A crucial phase is the validation of the co-constructed knowledge. It is the function of the teacher to verify the epistemological correctness of the constructed knowledge, ensuring that it doesn't contradict the existing knowledge. Such a contradiction could indicate either an error in the construction process, or the emergence of unexplored aspects of knowledge. In this perspective, the existing knowledge is no longer the object to be transmitted, but becomes a fixpoint of interaction. This teacher-student interaction needs to be parallel (symmetrical instead of parallel?), and synchronized with the interaction between the knowledge-knowing and knowledge-construction processes. It is not only cognitive but involves also the various actors in a holistic way via their embodied situatedness.

Laurilard describes this learning process as a recursive path between modulation and regulation, in which the student proposes first an idea and then the rules that recursively establish it. This recursive cycle inside the student is facilitated by the comparison with the teacher, with peers, and with the environment, which is composed of questions and answers, of tests and experiments that send intrinsic and extrinsic feedback. The student's initial goal-action-feedback takes place in the external environment, which includes both the physical and the social world, and can modulate the goals, actions, or concepts [18].

The concept of feedback plays a very important role as it allows the student to compare the knowledge gradually built, with other experiences or other findings that confirm and reject the results obtained. One of the limitations of many interactive processes is the lack of space for interaction and feedback. The absence of feedback often produces self-referentiality and satisfies only a closed system and that is not the result of interaction between a subject and its environment.

Specifically students with special needs require to continually confront their learning process with another external perspective, before they can develop the ability to self-regulate and can become external observers of their own work. The presence of an embodied feedback channel in form of a robot could become a fundamental support for learning processes. With the help of such robots the regulation of the learning process could be focused not only on cognitive results, but also on methods, timing, attention, participation.

3 Social Robotics

The goal of social robotics is the integration of robotic agents into mixed human robot ecologies, in which they interact in close physical proximity with humans. In order to achieve this, the interaction with this type of robot needs to be intuitive, flexible and pleasant, and the technology has to function reliably in complex social environments. Ideally the social robots will be able to identify the goals, intention and emotional states of their human counterparts and react to this sensory input appropriately. These reactions should be expressed verbally and non-verbally, and all aspects of the interaction between the humans and the robots should be synchronized.

This is of course still largely a vision for the future, but despite all the technical problems robotics - and specifically social robotics - is facing, great advancements have been made in this direction in the last 20 years and robots are slowly moving into human social spaces. Social robots have already been successfully implemented in different, yet very specific areas (Table 1). Various robots have been used in Robot Assisted Therapy (RAT) specifically for children with special needs [19, 20] and in elderly care [21].

Table 1. Examples of social robots used in different fields.

ROBOT	Embodiment characteristics	Field of application
KASPAR	Humanoid, minimally expressive child-sized robot with a static body	RAT for children with ASD and ADHD
PEPPER	Humanoid with cartoon-like characteristic	Entertainment, Research
PARO	Zoomorphic, comic-like	Elderly care, Dementia care
NAO	Small humanoid, comic-like	Education, RAT

What these applications have in common is that the robot has the function of a *social mediator*. It is used explicitly in the therapeutic process to connect patience with each other or with their care personal. The focus on the robots' role as social connectors is very important, since one of the fears is that robots will be used to substitute human contact. The objection usually is that humans, and specifically vulnerable individuals like children and the elderly, need other humans to have meaningful social interaction, and that interaction with social robot, which in the eyes of the critics only "fake" for example emotions and internal states, would unavoidably lead to a de-humanization of society. However, since the goal of an enactive social robotics approach is the connection of people rather than their separation this fundamental objection is not only unfeasible, but also carries inherently the danger of creating "human-like" machines, which people feel entitled to mistreat [22].

In general it can be said, that social robots are intended to have the capacity to facilitate human attachment towards them, to influence human behavior and emotional states, and to provide psychological support. This makes this technology a prime candidate for assigning to it a key position in the enactive didactics approach.

4 Enactive Robot Assisted Didactics (ERAD)

Due to the robots' specific properties – i.e., being embodied agents capable of expressing a multitude of social interaction behaviors – these artifacts have the potential to increase the complexity of interactions between the triad involved in enactive didactic process (teacher, student, and context) by reinforcing its reticular characteristics. Robots have the potential to become the gateway, which enables the students to interact with their environment and teachers on their own terms. The robot would become more than a mere tool for the achievement of the goals of the teacher. It would become the intermediary between the three elements involved in the teaching process. For example, in the case of special needs education, robots can help children with different types of disabilities to construct their own social reality, which will enable them to constitute new social interaction rules with other social agents (humans) in their environment. In the case of typically developing children, robots can be used to support learning processes related to kinesthetic and emotional aspects of social interaction [23].

What enables us to ascribe robots these roles are their unique properties within the social exchange process. Because they have a body equipped with sensors and effectors, they can react to their environment. The expression of their reactions depends on how we want them to be interpreted. They need certain degree of continuity in order to be meaningful, but beyond this their expressivity depends strongly on the aim of the context they are used in. It can be imagined that they can be used in different ways. Following von Foerster's line of argument, they could intervene when the didactic process becomes dualistic instead of enactive, i.e. they could at certain points ask "legitimate" questions. In this way they would help to shift the feeling of authority in the teacher "from knowing the answers" to the moderation of the enactive learning process. This would potentially make the teacher feel more at ease, and, in turn, improve the didactic process.

As briefly anticipated in Sect. 2, the robot could provide intrinsic feedback for the student. It could substitute the teacher in interpreting the knowledge constructed by the student through logic procedures implemented into it by the teacher. These procedures should however also be flexible in the interpretation of the knowledge produced by the student, and learn her cultural specifics and her mistakes, similarly to an intelligent tutoring system.

The robot would participate to define the ways in which the student thinks and generates models. The student would construct knowledge, to which the robot would provide direct feedback by performing either correctly or incorrectly.

In Fig. 2. we intend to illustrate two different situations - the current state of how robots are used in didactics, and a model for Enactive Robot Assisted Didactics. In comparison to Enactive Didactics (see Fig. 1.) the robot occupies the central of intermediator. The solid arrows indicate how robots like NAO [24] or BeeBot [25] are used at the moment in educational contexts. Their presence has an influence on the teacher, the student and the context, and their programmability allows them to be change by the educational context. However for this approach to become truly enactive the complexity of the structural coupling between all agents in the systems needs to be increased. The dashed arrows illustrate this increase. In order for the robot to become part of the dynamical changes within the learning process its reactive flexibility needs to be increased. It needs to be able to adjust to the behavior of the teacher and the student. This adjustment can be imagined in two different ways. The robot could either change its behaviors or its embodiment.

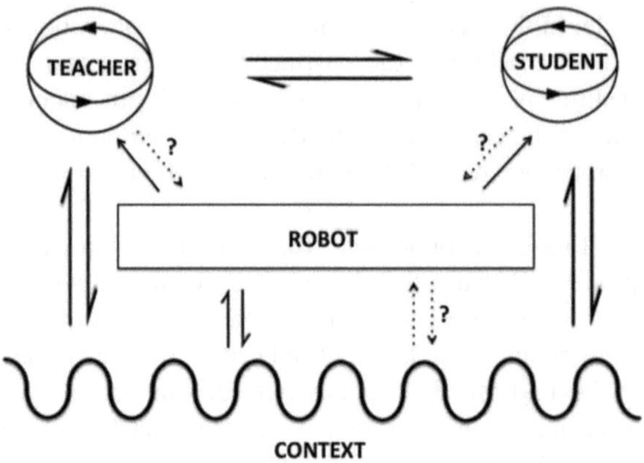

Fig. 2. ERAD - Structural coupling in current attempts of implementing an enactive approach in didactics, including a robot in the central intermediary function.

4.1 Shaping the Embodiment of the Robot

The only limitation potential shapes for robots can have is our imagination. We can think them as anthropomorphic, zoomorphic, or based on comic characters. Which embodiment a robot should have depends strongly not only on its function, but also on aesthetic considerations of its designer. Each embodiment has its advantages and disadvantages. Since the function of social robots is social communication, it stands to reason that they are equipped with communication channels that can be used to express interaction signals interpretable by humans. In many different studies it has been shown that giving the robot a face with animated features provides the humans a point of reference when addressing the robot (even if the robot hasn't for example a microphone its mouth or cameras in its eyes) [26], and on the other hand enables the robot to send signals that can be interpreted ad emotional states [27]. From the perspective of body language this also holds true. Humans do not only use their arms and hands to manipulate objects, but also for non-verbal communication [28]. Neurophysiological evidence from mirror-neuron system research shows, that watching a robot [29] perform certain actions initiates an activation pattern in the corresponding brain area in humans. This mechanism is thought to be linked to empathy [30], and it has been shown that, even on an expressive behavioral level, actions like contagious yawning can be triggered in humans by robots [31].

Based on this evidence introducing a robot with humanoid features like arms and a face in the classroom would prove advantageous for implementing the enactive approach we have illustrate in Fig. 2. Using a robot that has the ability to directly influence its human counterparts on a neuronal activation level could prove very effective during the teaching process.

There are however other factors that need to be taken in consideration when thinking about the embodiment of a robot for ERAD. In order to avoid the uncanny valley effect [32], it would be important at the current state of technology not to aim for an overly realistic humanoid robot, like for example the Geminoid series [33], but to find a way to integrate humanoid characteristics with for example cartoon-like features.

For the integration of the robot to be truly enactive, it would ideally be changeable. In terms of the ERAD approach this would mean that both the teacher and the student add new features to the robot based on the previous interaction they had with the system. We will discuss this idea in more detail in the last part of the paper, focusing on the potentialities of the maker movement.

4.2 Shaping the Behavior of the Robot

The second way how robot could be adjusted to the dynamics of the interaction it is involved is via its behavior. For the purpose of this paper we would categorize the social behavior of the robot into verbal and non-verbal.

Verbal behavior in this case consists of the robots ability to understand human verbalizations and to reply to them meaningfully. The problems social robotics is still facing in this area are related to speech recognition in noisy environments. These issues are gradually solved, however they are not yet in a state of being able to produce applications reliable enough for complex teaching situations.

Non-verbal behavior in general consists of movements of different parts of the body. Similar to the issues we are facing with the embodiment, we have to pay attention to the degree of human-likeness we want to achieve in movements and behaviors. Naturalistic movements in general create of higher level of interpretability and understandability. But even if the behaviors are very naturalistic, yet inconsistent and difficult to interpret, they still can generate the before mentioned uncanny valley effect. A good example for this is blinking, which also illustrates the conceptual closeness of embodiment and behavior. Human blinking signals engagement helps turn-taking during conversations and even signals internal states like the level of excitement or happiness. If we use a robot physically able to blink it is important to model the blinking appropriately using human physiological data. It has been shown, that humans rate a robot more intelligent and reliable when it exhibits naturalistic blinking patterns [34].

This example also shows that if robots with human-like features are used during social interaction processes, than these features create expectations of human-like behaviors in the human user. These expectations of human-likeness need to be fulfilled in order to make the interaction intuitive and comfortable. If on the other hand the expectations are not fulfilled, it is quite possible that the interaction is disrupted and the robot is rejected.

4.3 Summary

We propose an Enactive Robot Assisted Didactics Approach, in which robots function as catalysts between teacher, student and context in the construction of the classroom reality. Robots can have this function because they generate attention and expectations in both the teacher and the student, and are with their physical situatedness capable to influence and adjust the behavior of their human counterparts. In order for this process to be truly enactive, the robots need to be flexible in their interaction over time. This means their embodiment and behavior has to be adaptive. This opens various possible research trajectories involving the potentialities of the maker movement. It is important to train teachers in this technology, so they can use the robots to continuously 'learn how to learn'. This will in turn enable the student to continuously 'learn how to learn' in a dynamically changing environment.

5 ERAD and the Maker Movement

As we pointed out above there are different ways in which to shape the integration of robots in ERAD. To achieve a truly enactive didactic process the robots need to have the ability to influence all elements of the didactic triad – teacher, student and context, and all elements need influence to robot. With the advent of the maker movement era and the accompanying interconnectivity of its proponents the rapid prototyping and customization of the robot embodiment has come into the realm of the possible. People involved in the maker movement and in education can share their experience with different robotic platforms and discuss their needs concerning different scenarios.

We envision an accelerating process of exchange of specifications of robotic components that can be 3D printed according the affordances of specific didactic scenarios. Both teachers and students should be involved in this process. Such an embodiment design process would open an avenue of development that would be a first approximation of including robots in an enactive creation of the didactic reality. The possibility to share specifications of embodiment parts and rapidly produce them without being dependent on lengthy manufacturing processes would raise the level of empowerment of teachers and students and give a new meaning to the educational context. Education could finally move away from asking illegitimate questions towards an open participatory design process that would truly be enactive.

The other direction in robot design research that could potentially profit greatly from the possibilities intrinsic to the ideas of the maker movement is related to the need to personalize robots, specifically in educational settings and in contexts, in which the robots are used to teach children with special needs. It has been shown that the ability to personalize robots is specifically effective in settings that afford learning in young children [35]. But also in the above mentioned applications in RAT for children with ASD and children with conditions like Diabetes [36] the possibility to personalize robots has been proven to be useful. To make personalization of robots effective it is necessary to enable their users, in case of RAT for example the parents or therapists, the quickly understand the needs of their patients and incorporate them in new design features for the robot, which can be rapidly produced and shared in form of their specifications throughout the community. In extension this modus operandi would be very useful for the support of a truly inclusive approach to education support by the application of robots into the teaching process.

References

1. Varela, F., Thompson, E., Rosch, E.: The Embodied Mind: Cognitive Science and Human Experience. MIT Press, Cambridge (1991)
2. Rossi, P.G.: Didattica enattiva. Franco Angeli, Milano (2011)
3. Foerster, H.V.: Erkenntnistheorien und Selbstorganisation. Der Diskurs des Radikalen Konstruktivismus. Suhrkamp, Frankfurt (1987)
4. Bocchi, G., Damiano, L.: The enactive mind: an epistemological framework for radically embodied didactics. Educ. Sci. Soc. **4**(1) (2013)
5. Maturana, H., Varela, F.: Autopoiesis and Cognition. Reidel, Dordrecht (1980)
6. Wiener, N.: Cybernetics. Sci. Am. **179**(5), 14–19 (1948)
7. Pitts, W., McCulloch, W.: How we know universals: the perception of auditory and visual forms. Bull. Math. Biophys. **9**(1947), 127–147 (1947)
8. Duffy, B.R., Rooney, C., O'Hare, G.M., O'Donoghue, R.P.S.: What is a social robot? In: 10th Irish Conference on Artificial Intelligence & Cognitive Science. University College Cork, Ireland, 1–3 September 1999
9. Dumouchel, P., Damiano, D.: Living with Robots. Harvard University Press (2017)
10. Damiano, L., Dumouchel, P., Lehmann, H.: Towards human–robot affective co-evolution overcoming oppositions in constructing emotions and empathy. Int. J. Soc. Robot. **7**(1), 7–18 (2015)
11. Airenti, G.: The cognitive bases of anthropomorphism: from relatedness to empathy. Int. J. Soc. Robot. **7**(1), 117–127 (2015)
12. Broekens, J., Heerink, M., Rosendal, H.: Assistive social robots in elderly care: a review. Gerontechnology **8**(2), 94–103 (2009)
13. Feil-Seifer, D., Mataric, M.: Robot-assisted therapy for children with autism spectrum disorders. In: Proceedings of the 7th International Conference on Interaction Design and Children, pp. 49–52. ACM (2008)
14. Wada, K., Shibata, T., Musha, T., Kimura, S.: Robot therapy for elders affected by dementia. IEEE Eng. Med. Biol. Mag. **27**(4) (2008)
15. Maturana, H.R., Varela, F.J.: The Tree of Knowledge: The Biological Roots of Human Understanding. New Science Library/Shambhala Publications (1987)
16. Brown, J.S., Collins, A., Duguid, P.: Situated cognition and the culture of learning. Educ. Res. **18**(1), 32–42 (1989)
17. Damiano, E.: La mediazione didattica. Per una teoria dell'insegnamento: Per una teoria dell'insegnamento. FrancoAngeli (2013)
18. Laurillard, D.: Teaching as a design science. building pedagogical patterns for learning and technology (2012)
19. Dautenhahn, K., Nehaniv, C.L., Walters, M.L., Robins, B., Kose-Bagci, H., Mirza, N.A., Blow, M.: KASPAR–a minimally expressive humanoid robot for human–robot interaction research. Appl. Bionics Biomech. **6**(3–4), 369–397 (2009)
20. Vanderborght, B., Simut, R., Saldien, J., Pop, C., Rusu, A.S., Pintea, S., David, D.O.: Using the social robot probo as a social story telling agent for children with ASD. Interact. Stud. **13**(3), 348–372 (2012)
21. Bemelmans, R., Gelderblom, G.J., Jonker, P., De Witte, L.: Socially assistive robots in elderly care: a systematic review into effects and effectiveness. J. Am. Med. Directors Assoc. **13**(2), 114–120 (2012)
22. Damiano, L., Dumouchel, P.: Anthropomorphism in human-robot co-evolution. Front. Psychol. **9**, 468 (2018)

23. Barakova, E.I., Lourens, T.: Expressing and interpreting emotional movements in social games with robots. Pers. Ubiquitous Comput. **14**(5), 457–467 (2010)
24. Fridin, M.: Storytelling by a kindergarten social assistive robot: a tool for constructive learning in preschool education. Comput. Educ. **70**, 53–64 (2014)
25. Highfield, K., Mulligan, J., Hedberg, J.: Early mathematics learning through exploration with programmable toys. In: Proceedings of the Joint Meeting of PME, vol. 32, pp. 169–176 (2008)
26. Blow, M., Dautenhahn, K., Appleby, A., Nehaniv, C.L., Lee, D.: The art of designing robot faces: Dimensions for human-robot interaction. In: Proceedings of the 1st ACM SIGCHI/SIGART Conference on Human-Robot Interaction, pp. 331–332. ACM (2006)
27. Breazeal, C.: Emotion and sociable humanoid robots. Int. J. Hum Comput Stud. **59**(1–2), 119–155 (2003)
28. Argyle, M.: Non-verbal communication in human social interaction (1972)
29. Gazzola, V., Rizzolatti, G., Wicker, B., Keysers, C.: The anthropomorphic brain: the mirror neuron system responds to human and robotic actions. Neuroimage **35**(4), 1674–1684 (2007)
30. Gallese, V.: The shared manifold hypothesis. from mirror neurons to empathy. J. Conscious. Stud. **8**(5–6), 33–50 (2001)
31. Lehmann, H., Broz, F.: Contagious yawning in human-robot interaction. In: Companion of the 2018 ACM/IEEE International Conference on Human-Robot Interaction, pp. 173–174. ACM (2018)
32. Mori, M.: The uncanny valley. Energy **7**(4), 33–35 (1970)
33. Nishio, S., Ishiguro, H., Hagita, N.: Geminoid: teleoperated android of an existing person. In: Humanoid Robots: New Developments. InTech (2007)
34. Lehmann, H., Roncone, A., Pattacini, U., Metta, G.: Physiologically inspired blinking behavior for a humanoid robot. In: International Conference on Social Robotics, pp. 83–93. Springer, Cham (2016)
35. Baxter, P., Ashurst, E., Read, R., Kennedy, J., Belpaeme, T.: Robot education peers in a situated primary school study: personalisation promotes child learning. PLoS ONE **12**(5), e0178126 (2017)
36. Henkemans, O.A.B., Bierman, B.P., Janssen, J., Neerincx, M.A., Looije, R., van der Bosch, H., van der Giessen, J.A.: Using a robot to personalise health education for children with diabetes type 1: a pilot study. Patient Educ. Couns. **92**(2), 174–181 (2013)

Introducing the Making Culture in Teacher Education: The eCraft2Learn Project

Rene Alimisi[1], Dimitris Loukatos[1,2], Emmanouil Zoulias[1], and Dimitris Alimisis[1(✉)]

[1] EDUMOTIVA - European Lab for Educational Technology, Sparta, Greece
info@edumotiva.eu

[2] Department of Natural Resources Management and Agricultural Engineering, Agricultural University of Athens, Athens, Greece

Abstract. This paper presents the capacity building workshops for teachers that took place in the maker space established in Athens in the frame of the eCraft2Learn project. The aim of the workshops was to familiarise the teachers with the eCraft2Learn ecosystem that provides tools from robotics, DIY electronics, visual programming, 3D modelling and 3D printing in order to enable them to take on the role of the coach for their students in an environment inspired from the maker movement. The paper presents the training sessions and reports the trainees' feedback focused on the strong and the weak aspects of the eCraft2Learn tools, the challenges they faced, their understanding of the pedagogical ideas underpinning the eCraft2Learn initiative, and finally their confidence and preparedness for their role as coach.

Keywords: Educational robotics · eCraft2Learn project · Teacher training · Maker movement

1 Introduction

The maker movement has emerged recently in education with the great promise to democratize access to opportunities for learning by making and to the 21st century digital making technologies [1]. The maker movement challenges education systems to provide opportunities to access learning with cutting-edge making technologies for every child, for every citizen and not only for gifted people or science and technology majors [2]. From educational perspective, it has its roots in Papert's constructionism [3], the digital fabrication labs, the Do-it-Yourself (DIY) philosophy and envisions an education that will enable learners to make their own (digital or not) artefacts using the 21st century technologies [4].

Educational robotics in the makers' era is an essential part of this movement [5]. However, despite the positive effects, the spread of robotics and making technologies within school environments, the educational systems are not yet able to efficiently integrate robotics and maker movement into curriculum paths. In fact, new technologies are simply used to reinforce the old ways of teaching [6].

Being in line with this movement, the eCraft2Learn/H2020 research project, 2017–18 [7], researches, designs, pilots and validates an ecosystem based on digital

M. Moro et al. (Eds.): Edurobotics 2018, AISC 946, pp. 27–41, 2020.
https://doi.org/10.1007/978-3-030-18141-3_3

fabrication and making technologies for creating computer-supported artefacts. The project aims at reinforcing learning and teaching in science, technology, engineering, arts and math (STEAM) education and the development of 21st century skills that promote inclusion and employability for youth in the EU.

The eCraft2Learn project involves both teacher training and pilots with students. This paper focuses on the implementation of the capacity building workshops where the teachers/trainees experience first-hand the kinds of activities that are later implemented with students within the eCraft2Learn ecosystem. The first version of the workshops took place in Athens (September–October 2017) with the aim to familiarize the teachers with the eCraft2Learn methodology, technologies, tools and resources and to boost teachers undertaking the role of the coach in a maker space (i.e. the eCraft2Learn lab).

Section 2 describes the way the training workshops were implemented. Section 3 reports the feedback obtained by the trainees. Finally, Sect. 4 presents the conclusions and the experiences gained and indicates areas that are exploitable for future training courses.

2 Description of the Capacity Building Workshops

The trainees were introduced to the eCraft2Learn ecosystem in a similar way that they were expected to work with the students later in the pilots. General considerations that shaped the training plan included:

- Respecting teachers' time: the teachers could not devote a large amount of their time to the training. Thereby, our plan for the workshop had to consider the whole groups' time constraints and to come up with a realistic timeframe.
- Teachers were novices and needed support to feel confident: the participating teachers did not have previous experience in the eCraft2Learn tools/technologies. For most of them the idea of coaching was also something new. A challenge for the training team was to support them in exploring new tools and pedagogical practices without limiting the open nature of the workshop as it is engineered through the eCraft2Learn methodology that is put forward.
- Teachers needed to explore the different tools/technologies integrated into the eCraft2Learn ecosystem: it was equally important to scaffold teachers in exploring the different tools/technologies that support each stage of the eCraft2Learn methodology and to trigger their interest in doing increasingly more complex projects, combining different tools and entering different subject domains.

The training workshop was carried out with 20 teachers (10 male, 10 female) and was carried out in 5 sessions of 4 h each. The trainees came from different scientific fields, namely Science, Technology, Electronics, ICT, Graphic Arts. They had not been exposed to the eCraft2Learn tools/technologies before.

Briefly, the first session focused on the pedagogical ideas and the five stages of craft- and project-based methodology developed in the context of the project. The second, the third and the fourth session focused more on practical tasks using the eCraft2Learn technologies. The last session was dedicated to free experimentation with the introduced technologies and a discussion on pedagogical issues and on the forthcoming pilots with the students. The content of the five sessions and the methodology that was followed is presented shortly below.

Session 1

The session started with some ice breaking activities that aimed at bringing trainers and trainees closer and getting to know one-another. Time was then allocated to forming groups. In total, eight groups of two or three were formed. The participating teachers were then introduced to the five stages of the eCraft2Learn methodology. The methodology starts with the ideation stage where learners formulate their own ideas for the projects to develop by exploring the real world (stage 1). Then a planning stage follows where the learners exploring the resources available and needed for the realization of their ideas (stage 2). The learners then engage in a making process that includes brainstorming, iterative designs, trial and error and reflection on designs (stage 3) with the aim of building (using the available tools, technologies and resources) interactive artefacts and programming their behaviour (stage 4); finally, they share their projects with the class and the community (stage 5).

The research/training team presented and discussed in the context of group discussions the eCraft2Learn pedagogical methodology, the Maker Movement trend in education and the Do-It-Yourself culture. The pedagogical discussion lasted approximately two hours. The teachers documented their thoughts, ideas and concerns and all together were engaged in an interesting discussion that aimed at shedding light on the eCraft2Learn learning intervention that had just started.

From the pedagogical foundations, the groups of trainees smoothly entered the technical core of the eCraft2Learn learning ecosystem. The next step was the setting up of the work stations. The teachers were given a storage pack which included the necessary equipment and were encouraged to set up the Raspberry Pi (RPi) board [8], to connect the monitor, the keyboard and the mouse, to access the Wi-Fi internet connection and to explore the Raspbian environment. The idea behind this stage, which took place in groups, was to help teachers from the beginning to explore the main technologies that are included in the eCraft2Learn learning ecosystem. Engaging the teachers in the establishment of the work stations was also considered an important step towards helping them become adept in building similar work stations in other educational environments in the future.

After setting up their work stations, the trainees were introduced to the Arduino Uno boards, the Arduino software and the visual block programming tools [9] through simple tasks. The trainees were also encouraged to look for information online, to look at exploratory illustrations of the Arduino Uno board and work in groups to understand how the Arduino board works, how it is powered, how digital inputs and outputs are connected to the available pins and more. A document with guidelines had been designed to guide the trainees through several Arduino-related activities.

The trainees were supported to perform simple tasks that aimed at helping them understand what an Arduino board is, explore how to wire it, become familiar with the basic components of their Arduino kit (inputs, outputs, resistors, wires and breadboards, the solderless circuit prototyping boards that one can plug wires and components into etc.). learn how to connect the Arduino Uno to the Raspberry Pi board and become familiar with visual programming equivalents of the Arduino IDE environment like Ardublock [10] and Snap4Arduino [11] through practical tasks. The groups of trainees carried out most of these tasks going deeper into aspects that attracted their interest, learning from failures, exchanging ideas and thoughts, designing solutions on paper and asking for support and clarification from the trainers.

Session 2

In the 2nd session the trainees were encouraged to focus more on the visual programming environment of Snap4Arduino and to explore the visual blocks under the Control and Motion category and to create a script that makes the arrow sprite (in the stage of Snap4Arduino) move forward and backward. They further practiced how to save and export a project and how to make Arduino interactively communicate with the Snap4Arduino environment.

At a second stage the trainees were invited to connect a potentiometer in their Arduino board and to make the sprite in the stage of the Snap4Arduino environment turn according to the potentiometer settings. They were then asked to repeat this task by making an angle servo (that has been connected to the Arduino board) as well as a photoresistor turn certain angles according to the potentiometer settings. They then freely experimented with similar tasks, exploring further the interactive communication between the Snap4Arduino environment and their Arduino board. The session concluded with a discussion with the teachers aimed at gaining an insight into their thoughts, ideas for educational applications, concerns and needs.

Session 3

The third session was entitled "Introduction to Remote Control of ServoMotors Using RPi3 and Arduinos and Tablet Devices". The aim was to provide the trainees with opportunities to delve deeper into electronic circuit making, to further explore the servomotors and to prepare them to control their electronic artefacts remotely by using visual programming environments. The first activity of this session was the familiarisation with servomotors of moderate size so as to make them work using an Arduino Uno unit attached to the RPI3 unit. This setup requires the L293D driving circuit. Indeed, typical gear motors demand higher amperage to work than a LED, so we cannot directly connect them to the Arduino's digital outputs. In order to cover the gap, a dedicated circuit (usually the L293 chip) has to be used. As such, trainees could control a gear motor by connecting the digital pins 4 and 5 of the Arduino board to the L293D chip, making the rest of the wiring and using the Snap4Arduino environment to compose the necessary code blocks and test the results.

References to external resources were also made and the trainees were encouraged to continue their practice by consulting online resources i.e. on the Instructables platform, the Raspberry Pi documentation page and more.

In the context of this session, the trainees were also introduced to the GPIO functionality of the RPi3 and to the MIT Scratch environment [12] to control the GPIO pins of the RPi3 using broadcast messages. The steps that were undertaken involved: the activation of the GPIO server, the setting of a specific pin to act as an input or output and the assignment of a specific value to them. Building on this practice, the trainees made a LED to blink, and by using GPIO pins, they had to make a servo motor to rotate clockwise and counter-clockwise for some seconds.

By using the installed Virtual Network Computing (VNC) client application, under the "Internet" category of the basic RPi3 menu, trainees had access to the RPi3 unit of their neighbouring team and were able to repeat the same tests but now controlling the servomotors and the LEDs of their neighbouring team via the network.

The session continued with an introduction to the MIT App Inventor environment [13] which offers opportunities for remote control using tangible screens. The teachers were supported in exploring the environment and were challenged in programming a counter variable that changes values according to the buttons pressed.

The session concluded with a discussion with the teachers to set the plan in place for the forthcoming training sessions and to exchange ideas on the pedagogical application of the technologies introduced in the context of open projects for robotic artefact construction.

Session 4

This session was devoted to 3D modelling and printing tasks. First, the trainees were invited to brainstorm about the term 3D modelling and 3D printing and were introduced to key concepts included in the 3D printing. Open Educational Resources (explanatory pictures and videos) that demonstrated part of the 3D modelling and printing process were also used to familiarize the trainees with the practices underpinning 3D modelling and printing. Ultimately, the brainstorming brought also into focus the spectrum of the applications of 3D printing and the trainees further enriched the discussion based on information that they had recently come across, web articles that they had read, real life examples and more.

As in the previous sessions, a document with guidelines had been designed to support the trainees to perform simple 3D modelling tasks and become familiar with the TinkerCAD environment [14]. Initially, trainees were invited to design some simple 3D models and concurrently to explore the basic features of the TinkerCAD environment. The idea was to confidently work on simple tasks and then following their interests, to start exploring more advanced aspects in 3D modelling in TinkerCAD. The tasks that were carried out aimed at helping them

- understand how a shape is inserted into the work plane and is customized.
- explore the concepts of 'solids' and 'holes' and how the shapes (solids and/or holes) are combined/grouped together using the available tools.
- practically understand the concept of 'divide and conquer' which is behind the 3D modelling in TinkerCAD.
- practically analysing the final object from a geometrical perspective, dividing it in smaller units, which were 3D-modelled and grouped together in the end.
- experiment with the use of more than one work plane.
- understand how a project in TinkerCAD can be saved, imported to the slicing tool and transferred to the 3D printer.
- share their model in Thingiverse [15].

While the trainees were going through the tasks, they were advised to try out their own ideas or to freely extend the scenario of the task. Among the models that were designed were various models of houses, boats, buttons and rings. While half of the trainees were working on 3D modelling tasks, the rest were introduced to the different parts of the 3D printer and were asked to prepare a gcode file in Ultimaker Cura software [16] and to activate a 3D printing task with the support of a trainer (and vice versa).

Towards the end of the session, the teachers were encouraged to imagine how 3D modelling and printing can support the construction of a computer-supported artefact. This session lasted longer than had been originally scheduled; this was mainly since 3D printing tasks are time consuming and that the trainees were enthusiastically immersed in 3D modelling trying out multiple ideas.

Session 5

The first part of this session was a continuation of the 3rd session but introducing more sophisticated techniques for (remotely) controlling elements on the RPi3. The trainees had already learned in the previous sessions that by using the installed VNC client application (under the "Internet" category of the basic RPi menu) it is possible to connect to the RPi3 unit of a neighbouring team and work there using the Snap4Arduino environment or equivalent. This method, although straightforward, requires a large amount of computational power and bandwidth. In this session, they tried this task through the Snap4Arduino environment exploiting the simple HTTP requests creation mechanism the Snap4arduino has built in.

In summary, the main task for the trainees was to gain experience in techniques for the creation of simple HTTP requests, allowing for direct access/control of the GPIO pins of either local or remote RPis. To this end, after their introduction to the HTTP functionality in Snap4Arduino that is available through the "Sensing" category via the "http://" block they were invited to turn on and off a LED connected to the GPIO18 and the GND pins of the RPi board by pressing two buttons in the Snap4Arduino environment. To facilitate trainees' practical engagement in these tasks, a document with guidelines had been prepared for optional use.

The session was concluded with a pedagogical discussion about the craft- and project-based pedagogical approach that the eCraft2Learn project is putting forward. Several learning scenarios for indicative future activities were reviewed. The trainees were moved to envision how each stage can be implemented in the context of an eCraft2Learn activity/task, how the links to the authentic context of the real world could be enhanced, what technologies could be used to support each stage, what their role would be like and how students would be encouraged to take initiatives and to reflect on their learning.

The role of the teacher as a coach in the context of the forthcoming pilots at both pilot sites (formal and informal) was thoroughly discussed and the trainees were freely encouraged to reveal their thoughts and concerns. Emphasis was also placed on how a topic/lesson could be transformed into an eCraft2Learn task and how this could be designed and deployed taking on board students' interests and ideas.

In addition to the direct pedagogy-oriented interventions that took place in the first and the final session, the familiarization of the trainees with their envisioned role as coach of students was promoted throughout the course in two indirect ways. First, the trainees were encouraged in their practical tasks and projects to follow the eCraft2Learn pedagogical methodology, though adjusted to the short available training time. Second, the trainers acted less as instructors and mostly as coaches of the trainees helping discretely when it was needed and avoiding direct guidance. This way they provided a practical interpretation of the envisioned role of the coach to be used as an example for trainees to imitate.

3 Feedback Received from the Training Workshops

The teachers were asked to complete a questionnaire after the completion of the workshops. The set of questions put to the participating teachers appear below:

- Based on the experience gained during the training course, please mention the strong points of the eCraft2Learn technologies/tools that were introduced in this training workshop.
- Based on the experience gained during the training course, please mention the weak points of the eCraft2Learn technologies/tools that were introduced in this training workshop.
- Which are the main challenges that you encountered during this workshop?
- As a result of this training, do you feel more confident in your capacity to use the eCraft2Learn technologies as teacher/educator in the classroom or in other educational activities?
- Did the training course help you understand the pedagogical ideas underpinning an eCraft2Learn project?
- What changes should be made to enhance/improve this training workshop in the future?
- You will be invited to participate as a coach in the forthcoming pilots with the students. Please write your suggestions for the better implementation of the pilots both in a formal and informal educational setting here.

The comments made during the training workshops and the field notes kept by the trainers have been also documented and are critically presented. These comments arose in the context of periodically scheduled discussions with the group of teachers, during the training and after the end of the sessions. 16 teachers completed the questionnaire and all the teachers (20) participated in the discussions providing comments during the training workshops.

3.1 Results from the Questionnaire

The teachers' responses were grouped together under threads running through them and are summarized below.

Reference to Strong Points. Overall, a number of strengths were highlighted by participants including the rich variety of the technologies/tools that were explored during the workshop, the nature and the features of these technologies and tools, the pedagogical approach that was followed, the interactive discussions and the good communication with the trainers/project team. More precisely, the participants highlighted the following positive aspects related to the eCraft2Learn technical core:

- Free and open source technologies and tools.
- A great variety of technologies and tools that can support a wide spectrum of educational projects.
- Technologies that enable the connection between the digital and the physical world.
- Technologies that are low cost and can be easily extended into the school classrooms.
- Technological solutions that enable remote interaction among the machines.

Explicit reference was also made to specific technologies that are part of the eCraft2Learn ecosystem. More precisely, the trainees highlighted the flexibility in extending/advancing their project by connecting to the I/O Arduino pin multiple sensors and feedback components. The teachers highlighted also that visual programming environments ease the process for novices to step into the experience of programming without limiting opportunities to create more advanced programs. Their comments bring into focus Resnick's call (which draws on the ideas of epistemological pluralism [17] for learning environments that have not only a low floor (easy to get started with) but also a high ceiling (opportunities for increasingly complex explorations over time) and wide walls (a wide range of different explorations) [18].

The participants also highlighted the 3D printing technology, which was described as a hot topic in the world of technology with the potential to open new opportunities in the field of education. A lot of new knowledge was acquired and the opportunity to practically explore how software and hardware are brought together for the creation of a computer-based artefact was appreciated.

In their own words:

"The hands-on experience was the stronger point, along with the interesting technology (Arduino, 3d printing, etc.), that we don't usually come across. What is more, completing each task offers a sense of achievement in terms of making something from scratch" (Participant teacher)

"I see the familiarization with technologies and software that can clearly facilitate educational purposes as the strong aspect. Many tools were covered, and I tried them out and thanks to that I can make good suggestions to my students while they will be working on their constructions" (Participant teacher)

References to Weak Points and Recommendations. The issues raised by the participating teachers were related rather to the time that is required to master the new tools. Most of the teachers mentioned that time was a critical factor in becoming familiar with the eCraft2Learn tools. Based on their comments, the time that should be allocated cannot be easily pre-defined and is dependent on individuals' time constraints and needs. For instance, some teachers needed more time in becoming familiar with visual programming while others needed more time in the design of the electrical circuits. Very few teachers (2 out of 16) directly linked the free experimentation sessions with the lack of clear educational goals.

"... some confusion was in the free practice sessions. Perhaps more clear goals are needed for these sessions" (Participant teacher)

One teacher appears also to be skeptical about whether the eCraft2Learn initiative will be of interest to the students. Is he/she projecting his/her own anxiety on the students? How valid might such a concern be? These questions cannot be easily addressed at the moment but it worth revisiting after the completion of the pilots with the students.

"Perhaps, it may not fit into the interests of all students, because of the initial fear a student may have regarding technology" (Participant teacher)

In the light of this issue respondents made the following recommendations:

- A well designed and organised set of educational resources should be provided to the teachers
- More time should be allocated in the training that focuses on that specific set of tools
- Need for pre-developed simple examples for artefact construction
- Suggestions for 'homework-type tasks' to practice with
- Preparatory courses for novices in key thematic areas (i.e. in electronic circuit design)

In general, the participating teachers mentioned that they had gained an understanding of the pedagogical ideas underpinning the eCraft2Learn initiative. However, they stated that this occurred gradually as the training workshop was progressing and it is not easy for them to state the level of their understanding. The teachers also mentioned that regardless of the clear presentation of the pedagogical ideas, it is the actual implementation with the students that can confirm or not the level of their understanding.

"I made it in the end. First, I did not really get the whole coaching idea instead of being a traditional teacher" (Participant teacher).

"The pedagogical ideas underpinning the eCraft2Learn project were clear enough. Maybe we can tell more after working and planning activities with the students" (Participant teacher).

Challenges. The participating teachers were also asked to state the challenges that they had faced during the workshops. The documented challenges fall into three categories:

- challenges related to the eCraft2Learn technologies/tools.
- challenges related to the methodology that is put forward.
- challenges related to the time.

Regarding technology/tool- related challenges, some teachers (9 out of 16) mentioned that they mainly faced challenges with the DIY electronics (the construction of electrical circuits using the Arduino Uno board and the related components) and the programming. They closely related the challenges that they encountered to the absence of previous experience and pre-knowledge.

5 participants remarked as the main challenge the transfer of the eCraft2Learn learning methodology from theory to practice. In addition, the open nature of the forthcoming learning intervention and their new role as coaches was brought to the fore by the participating teachers.

"The main challenges were: A. to take responsibility of my learning, B. to explore, largely on my own, how to learn subjects I was not familiar with, C. I realized that I do not have to know all the details on the subject because later on -with the students- I will continue learning too" (Participant teacher)

"The biggest challenge is to accept the new spirit of teaching" (Participant teacher).

Last, 2 participants, whilst enjoying the content, had found the sessions quite intense owing to the range of tools practiced in each session.

Regarding Confidence & Recommendations. The teachers were also queried about their suggestions for the better implementation of the forthcoming pilots. 12 answers were provided, all of which related to the provision of guidelines for the teachers, worksheets for the students and supportive educational resources for both groups. Indicative comments follow:

"It would be useful to have access to organized educational materials for both trainers and trainees [..] These can support the development of the projects in the pilots with the students" (Participant teacher)

"Creation of possible scenarios for projects. These can help new ideas to emerge. Descriptive worksheets for the students [..]" (Participant teacher)

Based on their answers, organised educational resources related to the eCraft2Learn learning initiative and the components of the technological core can act as reference tools supporting their role as coaches.

Interestingly, among the comments collected, three describe a process that was very far from the eCraft2Learn concept. More precisely, three teachers requested pre-developed code and scripts well in advance for specific/predefined projects. This request could not be easily completed by the project team as it was in contradiction with the ideas underpinning the role of the teacher as a coach. Reflecting on their answers and the discussions that took place during the training sessions, it seems that they fear that they will be under-prepared and not be able to address students' queries or help them out properly. 'What if we cannot answer their questions or fail to support them?' In other words, it seems that some teachers tried to negotiate their role as coaches, approaching it from the more traditional and 'somehow safe' perspective of teaching. This type of 'negotiation' was expected given that the teachers were invited to go beyond their usual teaching practices and to exit their comfort zone. Some of them appeared to be more confident, other less confident but willing to take up the challenge and few others more reluctant. The statements below are indicative of the different levels of confidence that were observed in the groups.

"Maybe we need a more detailed (written) outline of an indicative activity; and (even better) we may try to do it in the lab before the workshop with the students. We can test it before hand, even if we are described as coaches. A good coach should know the subject and be so familiar with it, so that he can coach in a "cool" way and effectively"

"I do not feel sufficiently confident. However, I don't fear that I will not be able to coach a small team of students. It will be a challenge for me".

"Yes, I do feel confident. I really have to see it actually working with students in that special approach of learning - teaching"

From another point of view, the participating teachers' request for pre-developed code and scripts raises the issue of the extent to which the use of "ready-made" materials can be accepted. If the teachers ask for pre-developed scripts and explanations, then how can we convince them to inspire exploratory learning and making culture for their students? Learning can be influenced by emotional factors such as resistance, fear and the discomfort of not being as supportive as is expected (Pearson, 2010). The challenge for the project team was to boost the teachers' confidence and discretely support them to find their own way of coaching. "Half-baked" solutions that leave space for explorations may be the way to support their training in the beginning and to help them gradually build more confidence.

"I think that the workshop moved us smoothly to start thinking our role as students' counsellors, facilitators and partners. It also helps us get in the students' shoes and realize what it is expected from them. I enjoyed the discussions within the team and the collaborative spirit of this training" (Participant teacher).

3.2 Additional Comments Raised, and Observations Recorded During the Training

At the end of each session the training team was inviting the participating teachers to join an open group discussion in the context of which they could freely talk about their experiences, concerns, expectations and suggestions for improvements. The trainers were also initiated discussion about issues they may have observed, or comments raised during the training. The comments made during the sessions are documented and discussed below.

In the 1st training session, the trainees commenting on the 5-stages educational methodology suggested that the stages should not be used in a serial/linear way; the stages should be rather fluid and adaptable to students' needs and interests. Others added that students might start from different stages each time; for instance, a student might be in a situation where no initial ideas emerge and must start with the sharing step to form some ideas. Others suggested to encourage reflection on the different stages and recommended a "controlling stage" where students check their work and control their progress. Others focused on the interplay between ideation and planning and suggested that these stages are closely connected to each other. All agreed that sharing is a very important stage: students should share their projects at the end of each session or at the end of the pilots.

Another interesting remark initiated a lively discussion about the role of the student's apprenticeship and the initiation in learning process; the teacher expressed the opinion that the students learn better while working next to a 'wise' mentor. The discussion revolved around the question: 'how could the eCraft2Learn methodology offer such an initiation i.e. provide links to a community of experts?'

In the 2nd teacher training session the trainees were observed working with great interest. Some were curious to try out their own experimentations with Ardublock and Snap4Arduino programming tools going beyond the tasks proposed by the trainers. They were supportive to their peers; for instance, they were eager to help one another;

to share new knowledge (for example, after they had learnt a new tool). They seemed to enjoy their technical tasks and, when finishing a task, shared their achievements with the rest of the group.

They found that concurrently monitoring the Ardublock visual interface and Arduino IDE code helps to understand the code. They felt that making the circuits may be hard for students; a bigger breadboard or a breadboard properly marked to show how it is wired would be of help.

Regarding the training methodology, some trainees suggested that the training tasks should always be based on a scenario to make them more meaningful; they might then extend the scenario themselves depending on their interests.

Interestingly, in the 3rd teacher training session, though the training tasks in that session were more advanced, the trainees found the tasks easier than those of the two first sessions, possibly thanks to the experience they had gained and the skills they had already developed. This provides positive indications that our training methodology followed a stepwise approach. The trainees liked App Inventor and appreciated it as a good programming tool for kids. However, they warned to avoid giving too much information for children, since this might make the tasks difficult. Instead of this, they recommended providing proper information for children as 'food for thought' and foster inquiry learning, providing tools for the students to search for any information needed.

In the 4th teacher training session it was observed that the trainees were enthusiastically engaged in 3D modelling and printing activities, staying longer in the lab than scheduled. The familiarisation tasks were considered well-selected for novices and the session well-organised; in addition, they stated that they liked the intuitivity offered by the 3D modelling and printing tasks that they went through. They also seemed to start thinking about each technology and tool not separately but as a whole set that brings together technologies and tools that complement one another for the building of interactive artefacts.

> *"by today's 3D modelling and printing experience, things are getting glued together and completed, for example, we have learnt how to equip the artefacts with sensors, servos, logic and now how to give them a meaningful shape" (participant teacher)*

At the end of the 5th final session a plenary discussion was held for the overall evaluation of the course. Teachers appreciated they had made important progress during the course, acquiring knowledge and skills to deploy the eCraft2Learn technologies and the educational methodology. They recognised that the training methodology had offered a well-balanced mixture of "ready to use" and "left to discover" learning materials.

However, some teachers serving in the school education system were rather skeptical about the implementation in their school class. They felt insecure and that they would be inadequate in students' eyes because they do not have full knowledge of the tools. They asked for indicative solutions and detailed lesson plans for each project idea including objectives, tools, methodology, worksheets, and evaluation tools. Furthermore, they believed that school students do not have the basic knowledge of electronics, are not aware of the proposed tools and therefore would be not capable to finalize more than 2–3 of the proposed scenarios. They suggested to giving at least a worksheet to students with a short guide for the DIY electronic tools and the software to use in each learning session. Other suggestions included the provision of detailed materials for students and 'something to begin with'. These requests and concerns were also documented through their responses collected from the questionnaire.

4 Discussion and Conclusions

The skepticism and/or anxiety expressed by some of the teachers who serve the Greek public education system is well understood. The education system in Greece is mainly centralised and teachers are used to following directives and instructions coming from central educational authorities which rule the whole learning process in schools [19]. In addition to this, there are stereotypes for the role of teacher as an intellectual authority in the class who has ready answers for everything arising in the learning process. Therefore, the teaching methods are rather guided by the teacher and based only on handbooks approved by the ministry of education [19] not leaving much space for teacher and learner initiatives and resulting in predictable learning outcomes. Used to acting in such a framework, the eCraft2Learn teachers were suddenly invited to undertake a very different role, the role of the coach; to leave the initiative to students, to work in an unknown and challenging environment using completely new tools. All this can explain well and justify some frustration and anxiety obvious in some school teachers' attitudes. On the other hand, we noticed that trainees not serving in public schools (i.e. private school teachers, non-formal educators, young prospective teachers) were more open and confident in undertaking their new role, to adopt new tools and methods, and to be surprised by their students in the pilots to come. However, they expressed their apprehension that the equipment was not available at home to continue their training at their own private place and pace.

Based on the trainees' feedback, we have found that this entire training experience was useful for teachers in multiple ways. They experienced the physical participation in an authentic maker space (the eCraft2Learn lab) and were introduced to a broader concept of STEAM methods and tools; they conducted self-reflections on their experiences in the lab which helped to assimilate and adopt good practices; finally, they critically reflected upon their training endeavour, the problems and the challenges they faced.

The experiences that are documented here from our interaction with trainees can significantly inform future training workshops, limit traditional teaching practices and prepare teachers to take on the role of the coach going through the stages of defining the new role, doubting its applicability, negotiating its characteristics and nature, going through challenges, recognizing its educational potential, adopting it and keeping on improving it. This is what happened in the context of the eCraft2Learn capacity building workshops Most of the teachers got into the role of the coach with success while others understood its educational affordances and tried to find their own way of coaching.

The teacher training workshops invited trainees to dive into something they had not known how to do before, to explore a set of new tools and technologies and no matter the difficulties, the trainees took up the challenge and through their constant feedback showed us the way to support their educational journey. Their requests for educational resources and guidelines as well as tools for sharing were reviewed and informed our project work and indicative possible supportive means to accommodate trainees' preferences and allay their anxieties.

The teachers understood through personal experience that the aim is not the final artefact or the completed project but the underpinning learning process towards the completion that matters the most. Within the eCraft2Learn learning ecosystem, the teachers had the opportunity to see themselves leaving behind the role of the consumer/user and smoothly undertaking that of the maker. From that position, they can more easily stand by the students as co-designers and co-learners and it is more likely to calm their concerns and anxiety addressing an answer to the crucial question: *"what if I cannot answer the students' questions?"*

The challenges that were encountered were many, with the time factor being a common challenge. On the one hand the teachers' time constraints should be respected; on the other hand, the deployment of the eCraft2Learn methodology and the familiarisation with the eCraft2Learn technologies is a time-demanding task especially for novices. Future iterations of this workshop and other educational initiatives might seek to reconsider the balance between scheduled sessions and free open sessions for experimentation, perhaps organizing a pre-workshop for those in need of covering key concepts in STEAM or basic skills in assembling circuits, in using sensors and actuators and so on, in order to free up more time for practical skills training through their own projects.

To sum up, we see our work as contributing to teacher training in the STEM education field offering direct experiences and practical training in an authentic maker space environment. The work reported in this chapter was done in the frame of the eCraft2Learn project [7] that has resulted in the deliverable report "Capacity Building Workshops" [20] and combined with the other outputs of the eCraft2Learn project [7], will hopefully offer a full training package available for teachers and teacher trainers who wish to introduce digital fabrication and the maker movement in formal and informal education. This training package might be also integrated into academic STEM education courses for under- or post- graduate student-teachers to enrich the academic teaching practices with practical training in authentic maker spaces.

Acknowledgement. This research was supported by the eCraft2Learn project fund-ed by the European Union's Horizon 2020 Research and Innovation Action under Grant Agreement No 731345.

Disclaimer. This communication reflects the views only of the authors and the European Commission cannot be held responsible for any use which may be made of the information contained therein.

Statement on Ethical Approval for this Study. The ethical committee for the pilots of the eCraft2Learn project in Athens has received informed consent for this study signed by all the teachers who were involved in the pilots.

References

1. Blikstein, P.: Digital fabrication and 'making' in education: the democratization of invention. In: Walter-Herrmann, J., Büching, C. (eds.) FabLabs: of Machines, Makers and Inventors. Transcript Publishers, Bielefeld (2013)
2. Alimisis, D.: Teacher training in educational robotics: the ROBOESL project paradigm. J. Technol. Knowl. Learn. **24**(2), 279–290 (2018). https://lnkd.in/gzcVFaf
3. Papert, S., Harel, I.: Preface, situating constructionism. In: Harel, I., Papert, S. (Eds.) Constructionism, Research Reports and Essays, 1985–1990, p. 1, Norwood, NJ (1991)
4. Schon, S., Ebner, M., Kumar, S.: The maker movement implications from modern fabrication, new digital gadgets, and hacking for creative learning and teaching, In: Laia Canals, P.A.U. Education (Ed.) eLearning Papers Special edition, pp. 86–100 (2014)
5. Alimisis, D., Moro, M., Menegatti, E. (eds.): Educational robotics in the makers era. In: Advances in Intelligent Systems and Computing Book Series, vol. 560. Springer, Switzerland (2017)
6. Alimisis, D.: Educational robotics: open questions and new challenges. Themes Sci. Technol. Educ. **6**(1), 63–71 (2013)
7. eCraft2Learn/H2020 research project (2017–18). https://project.ecraft2learn.eu. Accessed 21 July 2018
8. Raspberry Pi. https://www.raspberrypi.org/. Accessed 29 July 2018
9. Arduino Uno. https://www.arduino.cc/. Accessed 29 July 2018
10. Ardublock. http://blog.ardublock.com. Accessed 29 July 2018
11. Snap4Arduino. http://snap4arduino.rocks/. Accessed 29 July 2018
12. Scratch. https://scratch.mit.edu. Accessed 29 July 2018
13. App Inventor. http://appinventor.mit.edu/explore. Accessed 29 July 2018
14. TinkerCAD. https://www.tinkercad.com/. Accessed 29 July 2018
15. Thingiverse. https://www.thingiverse.com/. Accessed 29 July 2018
16. Ultimaker Cura software. https://ultimaker.com/en/products/ultimaker-cura-software. Accessed 29 July 2018
17. Turkle, S., Papert, S.: Epistemological pluralism. Signs: J. Women Culture Soc. **16**(1), 128–157 (1990)
18. Resnick, M.: Sowing the seeds for a more creative society. Learn. Lead. Technol. **35**, 18–22 (2007)
19. Dimitropoulos, A., Kindi, V.: Accountability in Greek education, Country case study prepared for the 2017/8 Global Education Monitoring Report (2017). http://unesdoc.unesco.org/images/0025/002595/259533e.pdf. Accessed 18 June 2018
20. Deliverable 5.2-Report on the Capacity Building Workshops for Teachers. https://project.ecraft2learn.eu/wp-con-tent/uploads/2018/08/Deliverable_Report_Template_eCraft2Learn_D5.2_CapacityBuildingWorkshops_final_CSMCorrections.pdf. Accessed 10 June 2019

Pedagogy Behind Educational Robotics

Trans-constructionist Prototyping with a Rare Vintage

Micheál Ó Dúill(✉)

The British School of Sofia, Sofia, Bulgaria
logios.org@googlemail.com

Abstract. The constructionist philosophy has a flaw that makes it anti-educational. The flaw lies in the notion, cf. constructivism, that physical objects are information sources. This cannot be true. Physical entities must be mental compositions before they may become physical constructions. Papert's idea of an 'object to think with' is unsound: deriving concepts from percepts leads to error. The realm of number shows this. Thinking about objects, au contraire can be scientifically productive. For this, the notion of a 'physical entity' must encompass the symbolic representation of language. A logico-scientific evolution-based proof shows that perception offers a false view of reality; revealed and overcome through the thoughtful use of technology. On this foundation, a prototype design and technology course for primary school with an emphasis on robotics based on the LEGO® system was developed. Outlined and discussed, it is compared and contrasted with work from the MIT Media Lab.

Keywords: Construction · Computing · Objects · Evolution · Primary education

1 First Section

Castles on the beach, so solid when constructed with damp sand, dry out and erode as gentle breezes blow, and succumb completely to the advancing tide. Theories are upset not though didactic discourse but by evidence to contradict them. Philosophical cloaks offer protection from evidential forces. The appeal is to belief in the leader's wisdom, cf. religion. This can lead to anti-educational prescriptions. Constructionism and the maker movement need a firmer foundation than that offered [1].

Children in a post-Turing world, where the medium of education may now both assist and monitor their learning, need no longer be constrained to teaching method based on unresponsive pencil and paper. A better primary phase curriculum than that devised for the old medium is needed [2]. The constructivist/constructionist premise, that children construct their knowledge, is increasingly validated [3]. At issue is the role of objects in developing – or constraining – cognition and capability.

The answer this question is in the understanding that modern humans possess a unique and dangerous evolutionary adaptation (technicity [4]) that makes possible technology, thence science, art, and the deployment of entropy differentials massively beyond the biological range. Technicity is founded on information encoded in DNA over geological time and expressed in nervous systems.

M. Moro et al. (Eds.): Edurobotics 2018, AISC 946, pp. 45–56, 2020.
https://doi.org/10.1007/978-3-030-18141-3_4

2 Thinking with Objects

Papert, in his book "Mindstorms" [5] (now a LEGO Company registered trademark) took an object, the Turtle, and imbued it with mathetic properties. This microworld was a mathland that he believed children could explore using a capability called 'body syntonicity'. Children projected onto Turtle movements and verbalised in Turtle Talk. Turtle geometry [6] was associated with a programming language, Logo (LOGO), specifically devised for education. Papert's Mindstorms was to Logo as an epileptic seizure is for an individual sufferer: disruptive. A powerful computer language was cloaked with educational philosophy and degraded to Turtle Talk. Links to lambda calculus and the Turing machine were lost. Focus on a peripheral object rather than the language resulted in perceptualization at the expense of cognition. Control Lab Logo was called Logo-like by constructionists because it has no Turtle [7].

Post-Turtle, MIT focussed on the perceptible. For example the programmable brick (LEGO Mindstorms®) and its pictographic programming [8]. Scratch and Snap blocks-programming environments similarly privilege percept over concept. Though Mindstorms purported to be about providing a mathland that children might profitably explore, when Turtle and Turtle Talk were tested in the classroom they were found wanting. Its remnant is a button-operated robot that turns left, right and goes forward. Indeed, artefacts have insidiously inhibiting learning over generations. The constructionist flaw is in the difference between thinking-with and thinking-about.

2.1 Much Ado About Nothing

In the year twelve hundred and two Fibonacci brought a set of numerals developed in India to Europe from Arabia, including the cipher 0. With zero as place-holder, they were incompatible, practically and conceptually, with the tally-oriented Roman numeral system. Consider the numeral sequence '10'. Purportedly the counterpart of the Roman X, in binary it signifies two. There were now two incompatible conceptual models: (a) bundling into tens, versus (b) the working of a register machine (Fig. 1).

Fig. 1. Ten-bundle and zero-nine register as competing mental concepts of number.

If the notion that concepts are derived from percepts were true, then the Romans, users of loose-bead abaci, should have had the concept of zero as a number and placeholder. They did neither. The concept Zero flowed from its use in a decimal notation derived from language. Once Zero was written it became a publicly entity to be thought about and discussed. One conclusion is that, cognitively, the number system runs from zero to nine ant then increments an adjacent register. Failure to comprehend that nine, the capacity of the register, defines the decimal system leads to specious arguments against the linguistic foundation of number (e.g. the French 'quatre vingt douze' [9, 10]). For several centuries focus on the perceptible hindered acceptance of zero as a number, negative numbers, and the journey to the complex number plane via the square root of minus one, and more. It still hinders learning.

Zero has one mathematical property important for computing: division by zero is undefined. Failure to manage this properly leads to absurdities, e.g. MIT Scratch: Instead of saying that one divided by zero is undefined, the everyday answer of infinity is given. This implies non-termination. Eschewing error messages, Scratch lets blocks handle it in three different ways: (i) move gives the cat stage-fright and it rushes into the wings; (ii) repeat becomes an infinite loop; (iii) wait and turn treat it as zero; (iv) the letter picker treats it as one; and users are told the square root of minus one is not a number. This is hardly the mathland for which Papert pleaded.

Zero is fundamentally imperceptible. One bead can be made, no bead cannot. This negates the proposition that understanding flows from manipulation. Material like Dienes apparatus and the hundred-square, whilst aiding children to emulate a natural number calculator, inhibit that understanding of number needed for its application. Here is the fallacy at the heart of the constructionist philosophy.

2.2 Surface Thinking

Perceptualization similarly afflicts educational robotics. Consider the approach of the LEGO Company, a partner of the MIT Media Lab. Following the production in 1997 of the RCX programmable brick, its computing environment changed from the Logo text-based language to the pictographic NI LabView. The current EV3 and WeDo 2 kits still use it. For the latter Scratch also provides blocks. LEGO building instructions are pictorial, illustrating parts used and their sequential placing in the construction. Working at a perceptual level guarantees success. Pictographic programming attempts the same for coding. LEGO pictograms and Scratch blocks for WeDo 2 illustrate this. They draw attention is to the (arbitrary) direction of rotation of the motor (Fig. 2).

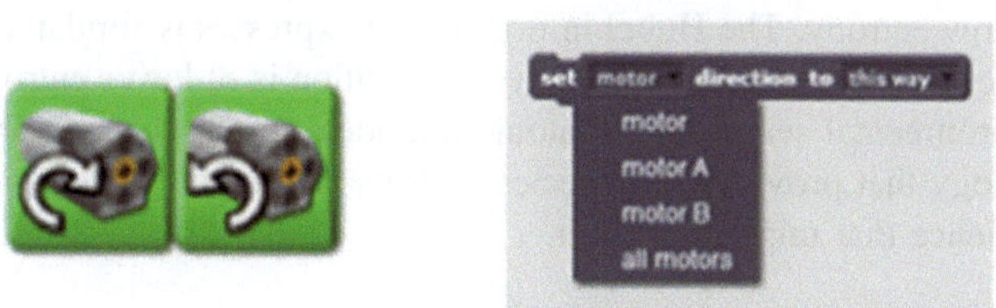

Fig. 2. LEGO WeDo NI LabView pictograms and associated Scratch motor control block.

LEGO do supply LED lamps, in pairs, designed for use as vehicle headlights. MIT do work with lamps using their Cricket board: designed for the RGB LED to enable light colour mixing, as on a TV screen. All these activities are perceptual, not cognitive. Despite working in a medium made possible by electromagnetism, there is no entry to the underlying science. This black box locks them into the immediacy of perception.

3 The Unreal World

The issues described so far are a direct consequence of the constructionist philosophy and its constructivist counterpart. Recognising that philosophies are not countered by science, it is nevertheless proposed to try to do so: by thinking about a pink sequin.

We perceive pink, be it a sequin on a party dress or a hedgerow dog rose, but we are not aware that colour has no reality. We know that the pattern of photons reflected from objects is transduced by three sets of cones, of different peak sensitivity, in the retina. The sensitivity is not arbitrary. Evolved by DNA and retained over geological time, it is part of the memory system: phyletic memory [11, 12]. Further into the brain, in Hubel blobs, is phyletic memory that creates colour from this data. Pink, mixed from red and white, matches no photon frequency. Form, colour, motion, tone, and all other sensations, are elemental phyletic memory [13, 14]. Brains construct a virtual representation of the world that is adaptive if not necessarily true. Colour vision is adaptive yet scientifically false. There are no 'objects' from which to extract number and other properties, rather the converse.

3.1 Perceptual Construction But Cognitive Composition

Phyletic information (memory) is the key to the perception-action cycle. Its false percepts enable organisms successfully to adapt in response to sense data. Perception does not obviously lead to a concept of pure colour. No other animal has shown the ability to seek out pure colour, as our ancestors did red ochre [15]. Geometric forms and pure tones, which are not environmentally experienced, exist only as phyletic memory. Future focussed prefrontal cortex composes potential action from memory. This opens up the possibility that this information might cognitively be compose into novel mental entities. Genetically, the most probable option is that the process of expansion of prefrontal cortex [16, 17] resulted in access to phyletic elemental memory, here designated Hubel information, in the human and only in the human. There is varying support for this evolutionary step and the consequent behavioural transition. Most telling is the capacity of modern humans to deploy entropy differentials well beyond the biological range. This implies a source of low entropy. DNA, a simple four base code, is low entropy. The Hubel information it expresses is similarly low entropy. Accessed directly by prefrontal cortex, any composition is of lower entropy than those based on environmental sense-data. Platonic-like ideal forms, they are the foundation of the technology that provides our species with its dangerous power over nature and the art and science that might control it.

3.2 Composition Precedes Construction

Pink sequins are composed from circular motion of a line about a point; a pink mixed from the colour palette; and shininess from elsewhere: all Hubel information. Creative cognitive composition of Hubel elements would better explain the character of childhood drawings than abstracted features: Children draw what they conceive [18]. When nine year-olds draw a picture of a LEGO construction they have built, they do not produce an architectural drawing showing the bricks used and how they combined them. They draw a 'cognitive' impression that hints at important features (Fig. 3).

Fig. 3. Drawing by nine year-old of a LEGO construction built as part of a spring project. There is a road with houses, cars, a dotted centre line, road signs and traffic lights. However all drawings are stereotypes, with no attempt to show that the original was LEGO. The roadworks are omitted and the traffic light given three lamps not two. This is creation not representation.

Fitters vs. Thinkers. The MIT Media Lab has a long association toy maker LEGO®. Success is the touchstone of the LEGO philosophy. No child who receives a LEGO set must fail to build it. Step-by-step pictorial instructions guarantee success. LEGO educational kits take the same error-avoidance approach. Behavioural objectives are successfully achieved. Understanding is uncertain. Bereft of instructions, untaught children will stack rather than bond bricks (Fig. 5a). Given only the bricks needed and a picture (Fig. 5b), university adults in computing also fail fully to bond the structure. They do not apply information the object-based approach presumes they extract. They compose an inadequate cognitive visualisation from internal information (Fig. 4).

Fig. 4. LEGO bricks stacked (L) and bonded (R) and house with bonding emphasized.

4 Story Time

Because primary-school children are still developing compositional capability, certain important ideas are introduced using stories. The electromagnetism that underpins our very existence and powers our culture is accessible this way. Language, recall, is an evolutionary adaptation predating our species. Children come to school with a fully developed mother tongue system and the capacity to learn others. Story telling is integral to all human cultures. Enter Ellie electron and Freddy photon.

Ellie (and her hard working friends) live in metal and can run around in it, but plastic keeps them in their playground. They only move when something attracts them – like a + at the other end of the battery. If they meet a lamp or motor on their way, they will make it work and go home tired. If they meet a LED lamp (a diode) they find that it has a door they must open. Here Ellie has a problem: she can push but not pull. So, depending on which way the door opens she is stuck outside – and can do nothing. If she can push the door open she will find Freddy photon, who is excited and wants to go out to play. But he can only go out to if he puts on the correct colour coat! Freddy and friends rush off in all directions, unless lined up with a laser.

Fig. 5. Ellie the electron and Freddy the photon, true fairies that work for us.

5 Prototyping with a Rare Vintage

Having established that concepts are not extracted from perception but are a product of our capacity cognitively to compose Hubel information, and so control the physical world, we are in a position to consider educational robotics environments.

5.1 Curriculum and Method Matters

Absent from the constructionist world-view is any concept of curriculum. This is not so for the schoolroom. Literacy, numeracy, and science are usually the curricula core. Thirty years after the computer came to primary school, computing was added.

Teaching method universally continues to presume pencil and paper as the primary educational medium. Use of computers in early literacy and numeracy is proscribed cf. Socrates and writing. Yet the register machine is faithful to Indian numerals and complements language. Zero is accessible, providing access to the trans-perceptual world of cognitive mathematics. In literacy the benefits of keyboard and screen with neat writing, spelling and grammar check are perceived as cheating. A medium with novel possibilities, its use clashes with beliefs based on its precursor. Although transition to Turing teaching [2] is a distant prospect, small subversions are possible.

Turing Prototyping. One safe haven for Turing enthusiasts is provided by building models of the real world that are meaningful to children and incorporate controllable elements. One such is the project-oriented curriculum developed by Ilieva [19, 20], based around elements from the LEGO Company. From 1995, the Control Lab 9 V system and (DOS) Logo software was used. Secondary school oriented, it had widgets and project pages; not useful in primary education. A (now rare) Windows version was released in 1996 (Fig. 7). It has proved to be excellent for prototyping an approach to primary school robotics that is congruent with and embedded in the developmental curriculum provided for children of this age.

The construction and computing curriculum runs from first to fourth grade. The method is project based with children working in teams. Four whole-school projects are built: for autumn, winter, spring and summer. These reflect the season, its flora and fauna, and any festivals. The aim is both to support the curriculum and develop analytical observation of the world around. Thus 'depth of perception' is improved by modelling entities in their world under the "Build and Animate YOUR WORLD" banner. On the cognitive plane, children's drawings in this developmental phase point to the need to represent critical features rather than realism. Hubel primary opponent pairs: red/green; blue/yellow; black/white; are ideally served by the basic LEGO hues.

The key form is the venerable brick, of two and one button width and varying length. (LEGO bricks have only two odd numbered lengths: one and three). These are supplemented by one third thick plates; slopes; beams; axles; wheels; gears; pulleys; pins; and a range of modified forms.

The 9 V electrical system is two-wire, carrying power and signal. The unique design of cable connectors and electric plates allow for right-angle connection.

The ungeared motor, used with gears and pulleys, has proved pedagogically the most versatile; similarly, the lamp with a single side light and through connection.

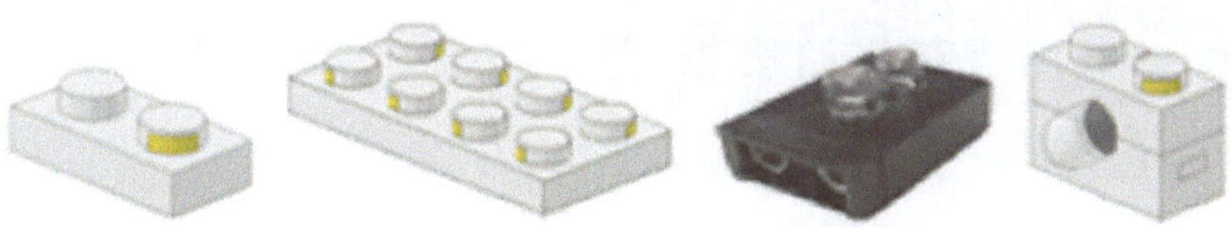

Fig. 6. LEGO 9 v electric plates from the 1990s. A red/green traffic light is built on a black 2-plate. Plus, a modifiable lamp with through connection and single side light.

Modernisation. Lamps are capable of having the original filament bulb replaced by a 3 mm LED. Additional electronic elements, e.g. solar panel; laser; speaker/amplifier; mp3 player; etc. may be built on an electric plate (Fig. 6). Though never included in a LEGO educational set, these plates are the key to adding more recent innovations. They are also used to supply electricity to street lamps and houses, from e.g. a battery box, simulating real world distribution. A 4 × 2 plate may be split to provide a 2 × 2 base. This vintage LEGO is relatively open, extremely versatile, and still available.

Subversion. This occurs in both the literacy and numeracy domains. Children type rather than write. (Mouse use is minimised and keys used where possible). Write the word 'on' on a piece of paper and nothing happens; make an error and wait for the teacher; punctuation has no visible effect. In the Turing world of Control Lab Logo all these have observable effects. In number, zero and the number line come alive.

5.2 The System in Outline

The key to the value of this Logo implementation is a relatively simple user interface and the possibility to write procedures that provide words meaningful in context. The LEGO Interface B is essentially as shown on the screen, with the exception that the buttons on the Output are lamps that indicate the positive terminal. Input indicators on the interface become text monitors on-screen. These show values from temperature, touch, light, and angle sensors. Outputs operate motors, lamps, and sounders. Output A is active by default. The central black 9v test contact is always-on. Communication with the computer is via an RS232 (serial) cable. Power is from an adaptor. The cable connectors and electric plates are an ingenious LEGO-brick design. Lamps are easily converted to LED operation. All is compatible with standard and technic LEGO.

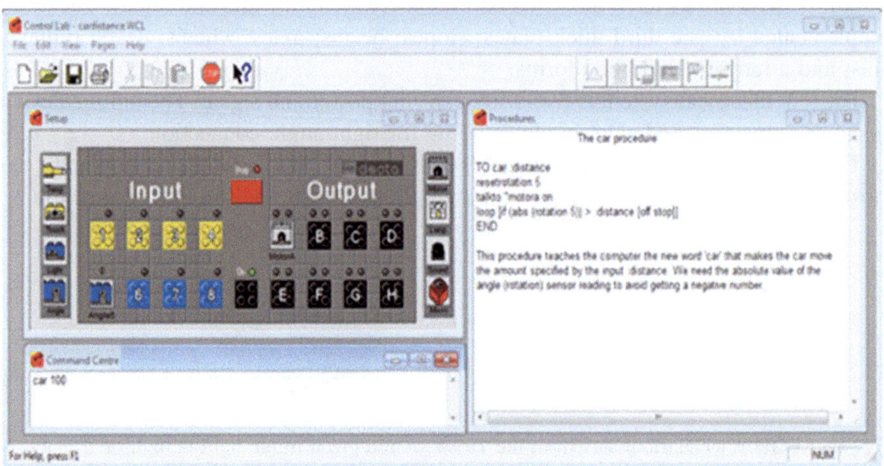

Fig. 7. The Control Lab user interface: command centre, setup and procedures pages.

By using the on-screen buttons, the children begin to understand that polarity (known from batteries and LEGO switches) affects how elements work: lamps flash or steady; motors rotate opposite ways; sound elements change tone. LEDs are diodes, so only work one way round; and there is a range of colours and types including bi-colour. Using a battery box or the test contact, a lamp can change from green to red by simply turning it around. Enter Ellie and Freddy and their adventures.

After working with the mouse and interface, and learning the keyboard, the children start to animate by writing code (linguistic-visual programming). The first program is 'cc' [ENTER] to erase the Welcome message from the Command Centre. Writing in the Command Centre may be reused by using the keyboard arrows to move up and down. With a bi-colour LED on Output A, typing 'on' turns it on, and 'off' off. For polarity we use 'setleft' and 'setright' (indicators light up appropriately). The 'rd' command will reverse the direction of the electric current, and thence its effect. It is necessary to be specific about elements attached when programming. A picture is dragged onto an output or input. Its default name then appears below it (see Fig. 7). The command 'talkto "motora setpower 2 on for 1 + random 50' will run a motor on low power for up to five seconds. Note: we call the double quote "rabbit ears". If a typing or coding error is made, feedback is given. Children receive a "blue message": e.g. onfor does not like 0 as an input, results from using 'random' without adding one.

5.3 Language

Computer language is more than coding; it is a means of expression. Logo, though English-like, has its own grammar, vocabulary, and punctuation. Translating the core vocabulary in not advisable and can be counterproductive. For instance, when the new version of Scratch lacked a Bulgarian translation it caused a hiatus in their education system. Unlike pictograms, blocks, or dialogue boxes, text is linguistically extendible and allows multiple levels of description. Lowest is the actions of LEGO elements, then components of the situation, followed by their behaviour in that situation. This method provides children with words meaningful in the context, e.g., 'red on for seconds 20,' when modelling a set of traffic lights, where 'red' is defined as 'talkto "lampa.' Plain text and procedure definitions, delimited by TO and END, may be mixed on the Procedures Page; so there is no need for semicolon prefixed comments.

5.4 Mathematics

The LEGO system with bricks, gears, links, etc. is full of number, shape and space. Children have fun turning a 2×2 block and finding it changes from square to diamond but have trouble working out that a three and a two block will fit in a five space.

Coding adds an educationally new dimension: active number with new challenges. Zero, is badly handled in most implementations. Children programming windmills use 'onfor', 'wait', and 'random'. Turning something on or waiting for zero time is seen as nonsense by the children, and Control Lab bans it. Random can deliver zero. To avoid an error they need to add one to the output of random, an important idea.

When the angle sensor is used to measure distance it delivers positive and negative numbers depending on the way it rotates. It introduces the negative side of the number line but the project requires only unsigned numbers. The sign may be removed using the 'abs' (absolute) operator, making the conditional for distance covered simpler and introducing an important number notion beyond those of correct computation.

5.5 Science

Electromagnetism is fundamental to existence. Children need to know that direction of current produces different results, e.g. direction of rotation. LEDs offer an entrée to photons. Bi-colour LEDs illustrate the effect of polarity and can be used in vehicles, changing from red to white according to direction of motion, and for traffic-lights. Photons are not normally visible and infrared lamps illustrate this.

5.6 Computing

Computer language is language; a second language. With Logo's first class procedure system writers can express complex operations in terms that are both closer to natural language and relevant to the situation modelled. The conditional used with sensors; the loop to detect a button-press; and launching independent processes e.g. doorbells in a street introduce important robotics ideas. The computational mind also introduces annoyances such as the inability to use a hyphen or a numeral as a character.

6 Curriculum Practice

The full set of projects from each grade is in the school documentation [21]. Illustrated below, with coding as appropriate, are grade-related elements that support aspects of the core curriculum, language, mathematics and science, in a manner appropriate to the world of robotics; all with a Turing twist and "small and beautiful."

First Grade. Road-works traffic-lights (lamp shown in Fig. 6) are modelled first using a (LEGO) battery box and polarity switch moved left and right. Command Centre code mirrors this using predefined procedures and outputs A and B:

```
loop [both.lights on setright wait 50 light1 setleft wait 200 light1 setright wait 50 light2 setleft
wait 200].
```

Second Grade. Simple three-colour traffic-lights are modelled. Lamps are assembled, taking care of electrical insulation, and coded using procedures written by the children, e.g. 'red.light' is 'talkto "lampa'. The overall super-procedure will then read:

```
red.light on wait seconds 25 amber.light on wait seconds 5 etc.
```

Project two uses motors, gears, and pulleys to simulate a windmill. Code introduces 'setpower' and 'onfor' used with 'random' and 'rd' to make it "dance" at varying speeds for varying times. The program for this raises the issue of zero (output by random to onfor) in computing. Using a bicolour LED it is shown that the motor is a generator.

Third Grade. A street of houses with touch-sensor doorbells introduces conditionals; launching independent processes; and revises loops. This is followed by programming Morse Code which, using a light sensor, laser, and teacher provided procedures, sends a message from one computer to another. It introduces list processing and ASCII code (to deal with the space character). This is followed by modelling the school environs and the new Bulgarian traffic-lights with flashing green and pedestrian phase and button. Here the complexity is in the algorithm and no new vocabulary is introduced.

Fourth Grade. A train introduces chassis and bodywork, worm gearing, and distance measurement with the angle sensor (and absolute value). For Christmas it is the Santa Express. Spring sees police action. A street of houses is equipped with burglar alarms, touch or light sensor (invisibly activated by an infrared LED), which sound an alarm in the police station and send a police-car to the correct house. Later, with the other grades, they build a fun-fair with rides, stalls, disco (with temperature sensor AC), etc.

7 Discussion

Computing in education and its robotics aspect were faced with a fork in the road at the beginning of the 1980s. In addition to the need for children to become capable with the new medium, a project hobbled by educational conservatism, two routes were sign-posted. Both originated in Logo and thence the lambda calculus and Turing machine, the cognitive foundation of computing. The first was a linguistic path that focussed on the language. The second way was object-oriented, harking back to the work of Piaget, and centred on a computer peripheral: the Turtle. This crystallised in constructionist philosophy and the maker movement. Was the progressive fork taken?

Analysis of Zero exposes a failure of constructionists to grasp the linguistic nature of number. If objects mislead us mathematically, the question arises of how real objects are. Thinking about pink led to the conclusion that perceptual constructs are false, yet adaptive, representations of reality. Humans uniquely possess a dangerous adaptation that makes possible the deployment of entropy differentials beyond the biological range and the creation of low entropy technology, science, art, and writing. Constructionists' object focus and demotion of language is an educational cul-de-sac.

The path that admitted writing as a cognitively composed entity was not taken. The programming language Logo is such an entity. It opens a window onto thought. In the form of LEGO-Logo, it offered primary school children opportunity to model their world. Control Lab Logo (once referred to as LOGO-like as it was Turtleless) has offered an excellent prototyping platform for a quarter century. The environment is educationally sound. With the help of a story, underlying science and mathematics may be revealed. The Logo language enables children to write code that is situated. The LEGO hardware is sufficiently open for user development. The drawback is that the software, like later LEGO hardware e.g. WeDo, is a closed proprietary product.

Over a quarter century, two teachers used this stable construction-and-computing platform to develop a core-linked curriculum that takes children from brick bonding to modelling action. Despite being educationally superior to current offerings from LEGO, MIT and elsewhere, using vintage material is not optimal. The interface box is a design classic, as is the connector, but the RS232 interface is obsolete. LEGO lamps are modifiable to LED, but there is no facility to add new sensors. For primary school the Dacta software is over complex and lacks some desirable features.

For a modern, tablet oriented implementation, LogoWriter Robotics, with its page metaphor and keyboard orientation, offers a good starting point. A basic specification might be: Command centre, setup screen, and procedures page where text, procedure definitions and pictures may coexist, as in a school notebook; tools for teachers to extend it, e.g. by adding sensors, as they do with traditional materiel; core vocabulary focussed upon children and their curricular needs; text editing; + & − usable e.g. +2 and set + left; no abbreviations; numerals re-definable. Children perceive 'to' as the infinitive and include it in the Command Centre, one solution is capitalisation.

For a disempowered classroom teacher, such notions are a dream. Well maintained, the kit will last another decade, and novel active LEGO bricks may be devised; but it needs a team to bring it up to current (and future) standards – and a manufacturer to supply it on a commercial basis. Whilst the team is possible, the latter is unlikely.

References

1. Papert, S.: Situating constructionism. In: Papert, S., Harel, I. (eds.) Constructionism. Ablex Publishing Corporation, Norwood (1991)
2. Dúill, Ó., M.: Teaching: Traditional vs. Turing. In: Reynolds, N., Webb, M. (eds.) WCCE 2013, vol. 2, pp. 9–16. Copernicus University, Torun (2013)
3. Fuster, J.M.: The Prefrontal Cortex, 5th edn. Academic Press, Amsterdam (2015)
4. Ó Dúill, M.: (2014) Countering and composing the mind's lies: technicity, science and art. In: Futschek, G., Kynigos, C. (eds.) CONSTRUCTIONISM 2014, Vienna, pp. 434–443. OCG, Vienna (2014)
5. Papert, S.: Mindstorms. Harvester Press, Brighton (1980)
6. Abelson, H., diSessa, A.: Turtle Geometry. MIT Press, Cambridge (1980)
7. Hecht, T.: LEGO dacta: a way to learn programming in a LOGO-like language. In: EUROLOGO 1997. ELTE, Budapest (1997). https://eurologo.web.elte.hu/prog.htm. Accessed 14 Feb 2019
8. Erwin, B., Rogers, C., Cyr, M., Portsmore, M.: ROBOLAB™: designing a visual learning environment with LabVIEW and LEGO, Tufts University, Medford (1999) https://amper.ped.muni.cz/~fyzika/projekt/download/robolwin.doc. Accessed 15 Jan 2019
9. Marmasse, N., Bletsas, A., Marti, S.: Numerical mechanisms and children's concept of numbers The Media Laboratory. Massachusetts Institute of Technology, Cambridge (2000)
10. Clayson, J. E.: Personal Communication (2010)
11. Fuster, J.M., Bressler, S.L.: Cognit activation: a mechanism enabling temporal integration in working memory. Trends Cogn. Sci. **16**(4), 207–218 (2012)
12. Fuster, J.M.: Phyletic Memory. Personal communication (2018)
13. Hubel, D.H.: Eye, Brain, and Vision. Henry Holt and Company, New York (1995)
14. Hubel, D.H., Wiesel, T.N.: Receptive fields of single neurones in the cat's striate cortex. J. Physiol. **148**(3), 574–591 (1959)
15. McBrearty, S.: Down with the revolution. In: Mellars, P., Boyle, K., Bar-Yosef, O., Stringer, C. (eds.) Rethinking the Human Revolution, pp. 133–152. McDonald Institute for Archaeological Research, Cambridge. (2007)
16. Deacon, T.W.: The Symbolic Species. W. W. Norton and Company, London (1998)
17. Streidter, G.F.: Principles of Brain Evolution. Sinauer Associates, Sunderland (2005)
18. Cox, M.V.: Children's Drawings. Penguin, London (1992)
19. Ilieva, V.: LEGO and Logo in the primary school – a simple way of learning through creation. In: Clayson, J.E., Kalas, I. (eds.) CONSTRUCTIONISM 2010, Paris. Comenius University, Bratislava (2010)
20. Doyle, M.P., Ilieva, V.: Primary robotics: laying sound foundations. In: Reynolds, N., Webb, M. (eds.) WCCE 2013, TORUN, vol. 2, pp. 28–35. Copernicus University, Torun. (2013)
21. Doyle, M.P.: Build and Animate Your World. Unpublished curriculum document. British School of Sofia (2018)

Creativity Thinking Skills Promoted Through Educational Robotics

Nikleia Eteokleous[1(✉)], Efi Nisiforou[2], and Christos Christodoulou[1]

[1] Frederick University, Nicosia, Cyprus
n.eteokleous@frederick.ac.cy
[2] University of Nicosia, Nicosia, Cyprus

Abstract. Creativity is the considered as one of the five key 21st century competencies for the twenty-first century. This research paper aimed to enhance creativity to thirty-two school students' while interactivity in programming with different robotic sets or tools as part of an informal curricular program. Main goal of these interventions was to nurture primary school students to computer programming skills and develop their problem solving and creativity. Results exemplified students' personal traits in their creative thinking in specific components of creativity. The paper showed that the methodological development of the educational robotics curriculum strives to utilize creative skills of young students.

Keywords: Educational robotics · Creativity skills · Elementary students

1 Introduction and Theoretical Framework

In robotics interventions, children are actively engaged, they collaborate and contribute on meaningful learning outcomes (Educational Robotics is a growing field with numerous researchers to have endorsed Robotics as educational tools [1, 2]. Robotics in the classroom has taken a global momentum especially because of its positive contributions in the teaching of science, technology, engineering and mathematics (STEM). Previous studies integrated robotics as an effective teaching method in the educational processes. These studies highlighted that in order to transform and improve the educational environment, robotics activities need to be appropriately designed and henceforth implemented in the teaching and learning practices [3, 4]. Besides, [5] identified the potential contribution of the incorporation of robotics as educational tool in schools.

The learning theories which apply in robotics education are constructivism and constructionism [6]. Robotics education follows the constructionist educational approach developed by Papert [6–9]. Constructionist learning, known as "learning through design" is based on the idea that individuals learn better when they are engaged in building and manipulating artefacts that are significant to them [9].

Student higher-order thinking skills (such as application, synthesis, evaluation, problem solving, decision making, and scientific investigation) are facilitated and promoted through robotics integration [7–14]. Using robots for educational purposes allows the development of different personal abilities [15, 16] and the 21st century

M. Moro et al. (Eds.): Edurobotics 2018, AISC 946, pp. 57–68, 2020.
https://doi.org/10.1007/978-3-030-18141-3_5

learning skills [6, 17]. One of the main values of educational robotics is its potential to inspire curiosity and creativity in students. Creativity in educational robotics has been associated with the constructionist learning paradigm and the processes of building, programming and manipulating robotic platforms [18]. Other studies found that the employment of robots in education fosters creativity in students [6, 17, 19, 20].

The current research intervention evaluates the application of a pioneer educational robotics curriculum delivered, in a non-formal educational setting (primary and secondary school) for the development of students' creativity. It investigates the impact of robotics integration as a cognitive-learning tool for the development of students' creative thinking.

1.1 Purpose and Research Objectives

While the benefits for robotics programs in education are plentiful, the purpose of this study was to explore how the Robotics Academy program may foster and develop the student creative behavior. Henceforth, the research has a two-fold objective:

- To evaluate the effectiveness of the robotics educational curriculum for developing students' creative thinking.
- To examine students changes in creativity after the completion of the educational robotics interventions.

2 Creativity and 21st Century Skills

Creativity is defined as a process that can create both original and effective products [21, 22]. Creativity reflects the capacity to solve problems in novel ways [23]. Novelty and usefulness are often referred to as the standard definition of creativity, which was first introduced by [21]. Creativity, which is an important twenty-first century skill, is essential for learning, work and daily life [24, 25]. The 21st century skills have been outlined and described by various researchers and reports, and can be summarized as follows: communication, collaboration, critical thinking, problem solving, knowledge construction, creativity, innovation, self-directed learning, global citizenship and digital literacy. It is increasingly important that students; as the future citizens of the Information Society are equipped with various 21st century skills in order to survive in this digitalized, globalized, and interconnected rapid-changing society and hence, to succeed in their career aspirations. Additionally, the challenge emerges in response to how best to cultivate students' creativity in the twenty-first century. Technology plays a crucial role in assimilation of these skills. Among flourishing arrays of technologies, emerging technologies such as robotics provide challenges and opportunities to the learners to develop innovative ideas, disruptive thinking and higher order learning skills. Robotic has been used as an educational tool from K-12 to graduate college [26].

Stager used a constructionism approaches to implement three different activities [27]: ballerina, teddy bear and Phonograph. Each activity was implemented with diverse groups and robotic platforms with different time span. Although he did not create these activities to foster creativity, they show how through robotics children can use their imagination to come with solutions to problems that are relevant to them. Therefore, researchers tend to define creativity as the ability to come with ideas or products that are novel and useful [28].

3 Creativity and Robotics

The development of creativity skills is important because of the changes in the global competition and collaboration, the focus on service economy, as well as the information growth. Given the aforementioned, the workforce needs have changed, the job tasks and type of work are changing and consequently the required skills are changing. Having said the above, what is the context, the environment and the tools through which these skills can be developed? How technology can contribute to the development of students' creativity skills that are considered important characteristics for today's globalized, interconnected world? The Educational Robotics Curriculum developed by the Robotics Academy aims to embrace all the above under its innovative umbrella. The RA Programme is designed for primary and secondary school students with the intention of raising their thinking ability. It is based on Hu's Thinking Ability Structure Model (TASM) [29], Piaget's cognitive development theory, and Vygotsky's social construction theory. The RA program has activities for 1st grade to 7th grade students in primary and secondary schools. Every grade has its specific manual, each including on average 25 activities covering creative thinking components.

There is still a challenge on how to exploit the advantages of a non-formal educational robotics program to effectively enhance students' creative skills and promote STEM education.

4 Methodological Development of the Educational Robotics Curriculum

A problem-based robotics program for elementary school students, was developed with the objective of reaching the students early on to instill an interest in STEM disciplines. The findings reported in this investigation are from the first 3 years of the program, when in-depth, formative evaluation data were collected. Authentic learning [30] formed the theoretical basis for the design of the robotics programs, which offers students the opportunity to learn not only math and science by putting numerous learning subjects into practice through robotics design and programming. Program goals for student participants in the robotics program were to: (a) acquire knowledge in various disciplines (mathematics, science, literature, language, history, geography, etc.) (b) develop 21st century skills, (b) encourage and develop interest in STEM disciplines and beyond, (c) build the necessary technological and digital skills, and (d) provide the students the opportunity to experience the joy of authentic creative thinking and problem-solving.

The Robotics Academy (RA) at Frederick University Cyprus (http://akrob.frederick.ac.cy & https://www.facebook.com/AkadimiaRompotikis) launched in order to establish and stimulate robotics mainly to the Cypriot educational system and society. It is a research and educational unit that aims to promote and conduct research in the area of robotics education. Specifically, it investigates how to best integrate robotics in the educational practice as a cognitive-learning tool within the educational process and impact of robotics in the development of various skills. The activities of the academy are multidimensional. More precisely, it includes the following: Educational robotics courses, Educational Robotics Program/Visits for Schools, Professional Development Training sessions to educators, Research Activity, Frederick University Events, Collaborations with various educational and social organizations, Social Charity Events & Volunteering. Based on the aforementioned experience and the scientific results of the experiments along the classroom interventions the educational material was developed.

The RA program developed an innovative curriculum for students and educators in elementary and secondary school levels. The philosophy and the pedagogical framework developed by the Robotics Academy serve as the backbone of the design of the educational robotics curriculum. It focuses on two elements: (1) Learning by Playing: building and Programming robots, and where (2) Robotics as partners in Learning (*Examine, Explore and Discover through Construction and Programming).*

The main purpose of the project was to enhance science and math learning experiences through educational robotics for students who have not had this type of opportunity in their regular school setting. The project had multiple goals to support its main purpose. The primary goals were to develop and test curriculum, curriculum materials and supplemental resources using the robot, geared toward an inner-city public-school population. A secondary goal was to examine the use of practical applications for the technology within a non-traditional educational environment in order to anticipate technical difficulties in our implementation plan.

The study was composed of two stages: (a) Design of the educational robotics interventions; (b) Students level of creative thinking through the completion of the Torrance Test of Creative Thinking (TTCT). Besides, the methodology employed the pedagogical approach of constructionism theory [4, 7] where students can learn better by doing. The methodology employed views educational robotic technologies not as mere tools, but rather as cognitive and learning tools that can serve as potential vehicles of new ways of thinking in the educational practices arena.

The project had multiple goals to support its main purpose. The primary goals were to develop and test curriculum, curriculum materials and supplemental resources using the robot, geared toward an inner-city public-school population. A secondary goal was to examine the use of practical applications for the technology within a non-traditional educational environment in order to anticipate technical difficulties in our implementation plan.

The study was composed of two stages: (a) Design of the educational robotics interventions; (b) Students level of creative thinking through the completion of the Torrance Test of Creative Thinking (TTCT). Besides, the methodology employed the pedagogical approach of constructionism theory [4, 7] where students can learn better by doing. The methodology employed views educational robotic technologies not as mere tools, but rather as cognitive and learning tools that can serve as potential vehicles of new ways of thinking in the educational practices arena.

4.1 Participants

The sample consisted of a total number of thirty-two students (30 boys and 2 girls) aged 5–12 years old from kindergarten to 7th grade were enrolled in the educational robotics program. The overall mean age was 8,34 years (SD = 1,73 years) (See Table 1).

Table 1. Students' demographic characteristics

Age	Grade	Pre & Post measurements	
		Frequency (N)	Percentage (%)
5	Kindergarten	2	6.3
6	1st Grade	4	12.5
7	2nd Grade	3	9.4
8	3rd Grade	6	18.8
9	4th Grade	10	31.3
10	5th Grade	4	12.5
11	6th Grade	2	6.3
12	7th Grade	1	3.1
	Total	32	100

4.2 Educational Robotics Curriculum and Interventions

The courses start on October and finish in May every year. Students attend one-hour educational robotics course each week (36 h within an academic year). The academy accepts students from six years old (1st elementary graders). The educational curriculum is divided in six different levels in order to cover all elementary grades/ages. The activities for each age group differ. Throughout the curriculum, the students experience a great variety of educational robotics packages and programming platforms: Bee Bots, Blue Bots, Robot – Mouse, Pro-Bot, Kibo Robotics Kit, Lego Mindstorms NXT, Lego Mindstorms EV3, Lego WeDo 1.0, Lego WeDo 2.0, Botley Robotics Kit, Little Bits, Grove, Edison Robot EdCreate Constructors Kit, Engino, mbot, Meet Edison. The exercises of the curriculum were developed based on the aforementioned packages and are intertwined with the following: presentations, educational games, documentaries, interactive audiovisual material, hands on activities, unplugged activities, and use of interactive educational software.

The first parameter of the robotics curriculum is based on the relationship of Robotics and Constructivism [3, 14]. The learning theory that supports constructivism highlights the educational value of such exercises as the ones the integration of robotics in the educational practice can provide. Such exercises are based on the philosophy of "learning by constructing", giving students the opportunity to develop interactive "thinking objects" [4, 12, 31–39].

The second element of the pedagogical is based on the following approach/ philosophy: Robotics integration in the teaching and learning practice is defined as the use of robotics by students as a tool that enhances their learning experience and supports the achievement of specific learning objectives [9, 12, 40–42]. This approach is related to the learning with computers or computers as mindtools, initially introduced by Jonassen [43], where computers and overall technology is introduced as students' partners within the teaching and learning process.

The educational robotics curriculum aims to develop participants' knowledge and skills in robotics (theory), in constructing various robotics models and in programming the robots. Various educational robotics packages and visual programing platforms are used. Through the curriculum, the participants are engaged in hands-on, technology-based as well as unplugged activities related to robotics, based on the grounds of gamification, project, problem and inquiry-based learning. Specifically, the educational robotics curriculum includes presentations, educational games, documentary, rich audiovisual material, hands-on activities, technology-based (educational software & simulations as well as unplugged activities, interactive activities (building & developing robots).

The educational robotics curriculum employs innovative approaches, which trace on the learning theory of constructionism where students are expose to their own learning construction when engaging in the making of concrete artifacts. The curriculum includes six structured 60- to 90-min activities and culminating interdisciplinary projects.

4.3 Torrance Test of Creativity Thinking (TTCT)

A Pre and post activity creativity psychometric tests were used to collect quantitative information for capturing students' level of creative thinking. The Torrance Test of Creative Thinking (TTCT) is a well-known and widely used tool for measuring creativity [44–46]. TTCT was employed to measure individuals' level of creativity and the impact of educational robotics curriculum in relation to creativity skills. The Figural Form of the Torrance Tests of Creative Thinking, the pre-test and post-test (TTCT) [47] were employed. The TTCT was developed by Torrance [48] and it is one of the most commonly tools used to measure the five dimensions of creative ability (Fluency, Originality, Abstractness of titles, Elaboration and Resistance to premature closure) and 13 creative strengths. In addition, there are 13 creativity strengths that were added to the average creativity score, that are considered the more complete overall creativity assessment, which are called the creativity index. The 13 creativity strengths consist of: (a) emotional expressiveness, (b) storytelling articulateness, (c) movement or action, (d) expressiveness of titles, (e) synthesis of incomplete figures, (f) synthesis of the lines, (g) unusual visualization, (h) internal visualization, (i) extending or breaking boundaries, (j) humor, (k) richness of imagery, (l) colorfulness of imagery, and (m) fantasy.

The TTCT is available in two versions; the TTCT-Verbal and the TTCT-Figural, each one consists of Form A and B. For the requirements of this study, the TTCT-Figural Response Booklets A & B [48, 49] were administered and scored according to the guidelines in the instruction manual and scoring guide [48–50]. The tests are divided into 3 non-verbal activities: Picture Construction, Picture Completion, and Lines and Circles (repeated figures). The completion of the tests requires a total working time of 30 min with a 10-min to complete each activity. Fluency is the number of relevant responses a person provides given a problematic prompt; flexibility is the variety of solutions and how diverse they are; originality refers to the unique nature of ideas; and elaboration is the depth of the description given for each solution [47].

The pre-measurement took place in October 2016 and the post- measurement took place in May 2017. The students attended the Educational Robotics Courses from October to May 2017, a total of 36 lessons. For the quantitative analysis, the targeted sample was elementary students' population, thus delivering questionnaires to all students (44). The response rate was 73%, since 32 questionnaires were completed and returned.

Lessons are organized to gradually introduce more complex tasks, but also to show case different subject matters or problems that can be solved using robots or programming and foster creativity. The lessons are divided into several small tasks, giving students a sense of progress, and essentially gamifying the learning experience. Students need to use their creativity and communicate ideas with other student to find possible solutions. Gradually, different concepts from other subject matters are introduced and incorporated in lessons. During the programming interventions students were introduced to and even master high level programming concepts. Many lessons were focused on support of creativity process with productive outcomes such as: students built a robot that created different drawings or played. The educational robotics activities are integrated in an interdisciplinary way and allow the learner to engage in a collaborative and creative approach, integrating complex- activities which require problem solve and creative skills.

5 Results

SPSS was used to analyze the data gathered. Descriptive (frequencies, percentages, means, and standard deviations) and inferential statistics (paired-sample t-test) were conducted to compare students' pre-test and post-test creativity scores. The study statistically examined the relationship between the robotics and the development of creativity skills when the educational robotics curriculum is employed. A paired-samples t-test was conducted to examine if there was a difference between the students' pre-test and post-test scores in "TTCT-Figural", Form A and B. Table 2 presents the information of the groups that participated. This information includes age, number of gender, and the means of the students' scores in pre-test and post-test relating to the five dimensions of the creativity test (fluency, originality, abstractness of the titles, elaboration and resistance to premature closure). The comparison between the pre-test and the post-test showed statistical significant differences, at the $p < .05$ significance level, in 3 out of the 5 components, namely Originality, Elaboration and Resistance to premature closure. Besides, great increase was observed in the total index of the 13 creative strengths compared to their scores in the pre-test and post-test.

Specifically, statistical significant differences emerged between the pre- and the post-tests in regards to age for the creativity component of Elaboration (Mean = 61.03, SD = 11.67 and Mean = 67.19, SD = 8.84; respectively), $t(31) = 3.16$, $p = 0.004$, showing the positive impact of the educational robotics curriculum experience on students creative thinking. There was a significant difference at $p < .05$ level between the means of the scores the students performed in the pre-test (Mean = 70.10, SD = 22.46) and the post-test Mean = 84.81, SD = 9.25, respectively) with regards to age for the creativity dimension of Resistance to premature closure, $t(31) = -3.46$, $p = 0.002$.

Similar results revealed in regards to the parameter of grade. Specifically, statistical significant differences emerged between the pre- and the post-tests in regards to grade for the creativity component of Originality, (Mean = 61.03, SD = 11.67 and Mean = 67.19, SD = 8.84; respectively), $t(31) = 3.16$, $p = 0.004$. Additionally, statistical significant differences revealed for the parameter of Elaboration (Mean = 61.03, SD = 11.67 and Mean = 67.19, SD = 8.84; respectively), $t(31) = 3.16$, $p = 0.004$, showing the positive impact of the educational robotics curriculum experience on students creative thinking. There was a significant difference at $p < .05$ level between the means of the scores the students performed in the pre-test (Mean = 70.07, SD = 22.36) and the post-test (Mean = 83.94, SD = 9.07, respectively) with regards to age for the creativity dimension of Resistance to premature closure, $t(31) = -3.30$, $p = 0.002$.

Finally, there was a significant difference in the scores for pre-test (M = 8.69, SD = 3.65) and post-test (M = 10.81, SD = 2.42) conditions relating to creative strengths, $t(31) = 3.07$, $p = 0.004$. However, it should be noted that there were no statistically significant differences in regards to age and gender between the creativity subscale measures; Fluency, Originality and Abstractness of Titles.

Overall, the results of the study demonstrated that educational robotics can positively influence the development of students' creative skills. In addition, the analyses revealed that is possible to employ robotics within the educational practice within a well-designed educational robotics curriculum where students experience various hands-on, technology-based as well as unplugged activities.

Table 2. Descriptive statistics and paired sample t-test results for the five creativity components

Creativity components	Pre-test		Post-test			
	M	SD	M	SD	t	p
Age						
Fluency	92.78	30.40	100.60	17.58	−1.408	.169
Originality	106.31	18.08	112.00	20.68	−1.598	.120
Elaboration	61.03	11.67	67.19	8.84	3.160	.004*
Abstractness of Titles	68.16	40.54	62.41	38.67	.570	.573
Closure	70.10	22.46	84.81	9.25	−3.458	.002*
Grade						
Fluency	93.22	30.66	100.03	17.90	−1.217	.233
Originality	105.88	18.35	111.63	20.82	1.39	.012*
Elaboration	59.22	11.48	66.03	8.10	3.459	.002*
Abstractness of titles	67.34	68.16	61.84	37.68	.551	.585
Closure	70.07	22.36	83.94	9.07	−3.303	.002*
Creative strengths	**8.69**	**3.65**	**10.81**	**2.42**	**3.069**	**.004***

Note. * $p < .05$.

6 Conclusion and Further Work

Creativity is a skill that been recognized as essential element of problem solving and critical thinking. With the objective to understand if robotics could be used to foster creativity, a non-formal educational program was developed in the context of a makerspace for children between 6 to 12 years old in Cyprus. A total of 36 workshops were implemented in the robotics academy, with a total of thirty-two participants. Quantitative data were collected and analyzed. The results showed the great potential of integrating robotics as a cognitive-learning tool, promoting research in the field of educational robotics. Ongoing work aims to expand the number of participants and investigate how educational robotics can influence students' creative thinking in STEM and non-STEM related subjects on a range of demographic characteristics such as gender, age and grade. The ambition is to suggest a set of guidelines on how teachers can enhance students' 21st century skills through coding, making, tinkering and play in formal and non-formal learning settings. The study exemplified the necessity to further examine and define the appropriate learning pedagogies and teaching approaches to be employed in the educational robotics curriculum. Henceforth, the transformation and improvement of the educational practice will promote the development of creativity skills needed for future citizens. Finally, more data on the development of creativity and educational robotics was collected throughout the school year 2017–2018. The data are in the process of being analyzed, aiming to compare and contrast with last year's data.

References

1. Frangou, S., Papanikolaou, K., Aravecchia, L., Montel, L., Ionita, S., Arlegui, J., Monfalcon, S.: Representative examples of implementing educational robotics in school based on the constructivist approach. In: Workshop Proceedings of SIMPAR, pp. 54–65, November 2008
2. Glăveanu, V.P.: Paradigms in the study of creativity: introducing the perspective of cultural psychology. New Ideas. Psychol. **28**(1), 79–93 (2010)
3. Bauerle, A., Gallagher, M.: Toying with technology: bridging the gap between education and engineering. In: Crawford, C., et al. (Ed.) Proceedings of Society for Information Technology and Teacher Education International Conference 2003, pp. 3538–3541. AACE, Chesapeake, VA (2003)
4. Papert, S.: Mindstorms: Children, Computers, and Powerful Ideas, 2nd edn. BasicBooks, New York (1993)
5. Benitti, F.B.V.: Exploring the educational potential of robotics in schools: a systematic review. Comput. Educ. **58**(3), 978–988 (2012)
6. Alimisis, D.: Educational robotics: open questions and new challenges. Themes Sci. Technol. Educ. **6**(1), 63–71 (2013)
7. Papert, S.: Constructionism vs. instructionism. In: Proceedings from Japanese Educators Conference (1980). http://www.papert.org/articles/const_inst
8. Vygotsky, L.S.: Mind in Society: The Development of Higher Psychological Processes. Harvard university press, Cambridge (1980)

9. Eguchi, A.: What is educational robotics? theories behind it and practical implementation. In: Society for Information Technology & Teacher Education International Conference, pp. 4006–4014. Association for the Advancement of Computing in Education (AACE), March 2010
10. Miglino, O., Lund, H.H., Cardaci, M.: Robotics as an educational tool. J. Interact. Learn. Res. **10**(1), 25–47 (1999)
11. Resnick, M.: Playful learning and creative societies. Educ. Update **8**(6) (2003). http://web.media.mit.edu/wmres/papers/education-update.pdf
12. Bers, M.U., Ponte, I., Juelich, C., Viera, A., Schenker, J.: Teachers as designers: integrating robotics in early childhood education [Electronic Version]. Inf. Technol. Child. Educ. Ann. **2002**, 123–145 (2002)
13. Chambers, J., Carbonaro, M.: Designing, developing and implementing a course on LEGO robotics for technology teacher education. J. Technol. Teach. Educ. **11**(2), 209–241 (2003)
14. Williams, D., Ma, Y., Prejean, L.: A preliminary study exploring the use of fictional narrative in robotics activities. J. Comput. Math. Sci. Teach. **29**(1), 51–71 (2010)
15. Chen, G.D., Chang, C.W.: Using humanoid robots as instructional media in elementary language education. In: 2008 Second IEEE International Conference on Digital Games and Intelligent Toys Based Education, pp. 201–202. IEEE, November 2008
16. Miller, D.P., Nourbakhsh, I.R., Siegwart, R.: Robots for education. In: Springer Handbook of Robotics, pp. 1283–1301. Springer, Heidelberg (2008)
17. Khanlari, A.: Effects of robotics on 21st century skills. Eur. Sci. J. ESJ **9**(27), 27–36 (2013)
18. Zawieska, K., Duffy, B.R.: The social construction of creativity in educational robotics. In: Szewczyk, R., Zieliński, C., Kaliczyńska, M. (Eds.) Progress in Automation, Robotics and Measuring Techniques 2015, pp. 329–338. Springer, Heidelberg (2015)
19. Botelho, S.S., Braz, L.G., Rodrigues, R.N.: Exploring creativity and sociability with an accessible educational robotic kit. In: Proceedings of 3rd International Conference on Robotics in Education (RiE 2012), Prague, Czech Republic, pp. 55–60 (2012)
20. Eteokleous, N.: Employing educational robotics for the development of problem-based learning skills. Paper presented at the 2015 Annual Meeting of the American Educational Research Association, Chicago, IL (2015)
21. Stein, M.I.: Creativity and culture. J. Psychol. Interdisc. Appl. **36**(2), 311–322 (1953)
22. Runco, M.A., Jaeger, G.J.: The standard definition of creativity. Creativity Res. J. **24**(1), 92–96 (2012)
23. Razumnikova, O.M., Zagainova, A.U.: The significance of motor asymmetry and hemispheric characteristics for effective convergent and divergent thinking during information identification. J. Asymmetry **7**, 2 (2013)
24. Johnson, P.: The 21st century skills movement. Educ. Leadersh. **67**(1), 11 (2009)
25. Trilling, B., Fadel, C.: 21st Century Skills: Learning for Life in Our Times. Jossey-Bass, San Francisco (2009)
26. Matari, M.J.: Robotics education for all ages. In: Proceedings of AAAI Spring Symposium on Accessible, Hands-on AI and Robotics Education, March 2004
27. Stager, G.: A constructionist approach to teaching with robotics. In: Proceedings of Constructionism and Creativity Conference, Paris, France, pp. 16–20, August 2010
28. Batey, M.: The measurement of creativity: from definitional consensus to the introduction of a new heuristic framework. Creativity Res. J. **24**(1), 55–65 (2012)
29. Hu, W., Wu, B., Jia, X., Yi, X., Duan, C., Meyer, W., Kaufman, J.C.: Increasing students' scientific creativity: the "learn to think" intervention program. J. Creative Behav. **47**(1), 3–21 (2013)

30. Lombardi, M.M.: Authentic learning for the 21st century: an overview. EDUCAUSE Learning Initiative, Boulder, CO (2007). http://www.educause.edu/ir/library/pdf/ELI3009.pdf
31. Eteokleous, N.: Developing 21st century skills through an educational robotics curriculum. Paper presented at the 2016 Annual Meeting of the American Educational Research Association, Washington, DC (2016)
32. Kazakoff, E., Bers, M.: Programming in a robotics context in the kindergarten classroom: the impact on sequencing skills. J. Educ. Multimedia Hypermedia **21**(4), 371–391 (2012)
33. Kazakoff, E.R., Bers, M.U.: Put your robot in, put your robot out: sequencing through programming robots in early childhood. J. Educ. Comput. Res. **50**(4), 553–573 (2014)
34. Kafai, Y., Resnick, M.: Constructionism in Practice: Designing, Thinking, and Learning in a Digital World. Lawrence Erlbaum, Mahwah (1996)
35. Harel, I., Papert, S.: Constructionism. Ablex, Norwood (1991)
36. Papert, S.: Constructionism: A New Opportunity for Elementary Science Education. A MIT proposal to the National Science Foundation (1986)
37. Puntambekar, S., Kolodner, J.L.: Toward implementing distributed scaffolding: helping students learn science from design. J. Res. Sci. Teach. Teach. Educ. **42**(2), 185–217 (2005)
38. Sullivan, F.R., Moriarty, M.A.: Robotics and discovery learning: pedagogical beliefs, teacher practice, and technology integration. J. Technol. Teach. Educ. **17**(1), 109–142 (2009)
39. Sullivan, A., Bers, M.U.: Robotics in the early childhood classroom: learning outcomes from an 8-week robotics curriculum in pre-kindergarten through second grade. Int. J. Technol. **26**(1), 3–20 (2015)
40. Bernstein, D., Mutch-Jones, K., Cassidy, M., Hamner, E., Cross, J.: Robots and Romeo and Juliet: studying teacher integration of robotics into middle school curricula. Paper presented at the 2016 Annual Meeting of the American Educational Research Assosiation, Washington, DC (2016)
41. Eteokleous, N., Demetriou, A.Y., Stylianou, A.: The pedagogical framework for integrating robotics as an interdisciplinary learning – cognitive tool. In: Roselli, J., Gulick, E. (Eds.) Information and Communications Technology: New Research, pp. 141–158. Nova Science Publishers, Inc (2013)
42. Thomaz, S., Aglae, A., Fernandes, C., Pitta, R., Azevedo, S., Burlamaqui, A., et al.: RoboEduc: a pedagogical tool to support educational robotics. In: Proceedings of ASEE/IEEE Frontiers in Education Conference, San Antonio, TX (2009)
43. Jonassen, D.H.: Computer as Mindtools in Schools: Engaging Critical Thinking, 2nd edn. Prentice Hall, Colombus (1999)
44. Baer, J.: Divergent Thinking and Creativity: A Task-Specific Approach. Lawrence Erlbaum Associates, Hillsdale (1993)
45. Kim, K.H.: Can we trust creativity tests? a review of the torrance tests of creative thinking (TTCT). Creativity Res. J. **18**(1), 3–14 (2006)
46. Nisiforou, E.A.: Examining the association between users creative thinking and field dependence-independence cognitive style through eye movement components. In: Proceedings of the 2015 ACM SIGCHI Conference on Creativity and Cognition, pp. 205–208. ACM, June 2015
47. Torrance, E.P.: The Torrance Tests of Creative Thinking-Norms-Technical Manual Research Edition-Verbal Tests, Forms A and B- Figural Tests, Forms A and B. Personnel Press, Princeton (1974)

48. Torrance, E.P.: Torrance Tests of Creative Thinking: Technical Norms Manual. Personnel Press, Lexington (1966)
49. Torrance, E.P., Ball, O.E.: The torrance tests of creative thinking streamlined (revised) manual, figural A and B. In: Davis, G.A., Rimm, S.B. (eds.) Education of the Gifted and Talented. Allyn & Bacon, Boston (1998). Scholastic Testing Service, Bensenville, IL (1984)
50. Torrance, E.P., Ball, O.: Streamlined scoring and norms for figural form A and B. Georgia Studies of Creative Behavior, Athens (1978)

From Digital Fabrication to Meaningful Creations: Pedagogical Perspectives

Calkin Suero Montero[1(✉)], Christian Voigt[2], and Kati Mäkitalo[3]

[1] School of Educational Sciences and Psychology, University of Eastern Finland, Joensuu, Finland
calkins@uef.fi

[2] Zentrum für Soziale Innovation, Vienna, Austria
voigt@zsi.at

[3] Faculty of Education, University of Oulu, Oulu, Finland
kati.makitalo@oulu.fi

Abstract. Digital fabrication consists of a range of techniques used to create products from digital designs. For the development of the digital citizen, digital literacy is a fundamental skill that digital fabrication could assist in developing. However, the main question is how to support education systems in order to promote fundamental skills, since the history of educational technology shows that changing technology requires changes in teacher's skills, curricular foundations and technical infrastructures in general. In this paper, we look at pedagogical approaches that facilitate the intake of technologies to be used in educational contexts through digital fabrication. First, we present our design thinking based approach encompassing five stages: *ideation*, *planning*, *creation*, *programming* and *sharing,* to imbue digital fabrication with meaningful creations in education. The paper continues to present students' work and provides a summary feedback from teachers. We conclude by arguing that teaching approaches and learning stages are good to structure learning through digital fabrication ex-ante, and that there is a need to also support the experiential dimension on the teachers' side by acquiring a 'coach's mindset', following students' ideas.

Keywords: Technology innovation · Craft- and project-based pedagogy · Inquiry-based learning · Design thinking

1 Introduction

Digital fabrication allows to (re)create, tinker and develop artefacts using digital technologies such as 3D design, 3D printing and laser cutting [27]. In education, digital fabrication is also tied to the rise of the *maker movement*[1] and the spread of the *FabLab movement,*[2] both based on Papert's [18] constructionist agenda in order "to empower,

[1] *Maker culture* is a technology-based extension of the DIY culture encouraging novel technology applications or creation (see for instance [30]).

[2] A *Fablab* (fabrication lab) is low cost lab equipped with digital fabrication technologies [3].

M. Moro et al. (Eds.): Edurobotics 2018, AISC 946, pp. 69–82, 2020.
https://doi.org/10.1007/978-3-030-18141-3_6

to educate and to create 'almost everything'" (see Nunez [17] and Gershenfeld [8]). Similar educational objectives can be found much earlier in Montessori's quest to attune education to children's individual development stages, rather than assuming that children would attune to prescribed curriculum [9]. Empowering children in the sense that they could *steer* the machine rather than *following* the machine's instructions was then implemented by Papert's Turtle Graphics, using the educational programming language 'Logo' in the late 1960s. This work continued by bringing fabrication facilitates to school children as described in the work of Blikstein [3], and Blikstein and Krannich [4]. Such empowerment is not only useful to develop competences in science, technology, engineering and math but also to understand better the current digital society in which we live in and the self (see Schelhowe [21] as cited by Smith et al. [22]) as a digital citizen.

Moreover, digital fabrication in education has also been praised for fostering the development of 21st century digital citizen skills including creativity, teamwork and problem solving [1]. This shows the enormous potential that digital fabrication presents for enhancing educational activities. One problem that seems to appear with the inclusion of digital fabrication in education is, however, that school children and teachers alike may fall prey to only make predefined, risk-free and oversimplified exercises, where construction becomes a mere "push-button" activity [6]. This type of activities and exercises in the end might only diminish the learning potential that digital fabrication technology can actually afford. Blikstein points out this "keychain syndrome" [3], where students and teachers are only involved in producing simple aesthetic artefacts such as a keychain. Our work proposes to go beyond the creation of simple aesthetic objects to reap the potential that digital fabrication and *making* can bring to education: empowerment to create interesting and meaningful artefacts while applying the concepts that fuel inquiry based learning.

Therefore, in this paper, we look at how digital fabrication is deployed through a solid inquiry based, craft- and project-based pedagogical approach. The idea is to bring forward the importance of a pedagogical framework for teachers and educators when using digital fabrication in the classroom. We argue that the combination of a pedagogical approach with the endless learning opportunities afforded by the use of digital fabrication in the classroom will facilitate the inclusion of innovative technologies in education. Working on Papert's constructionist ideas, our aim is to facilitate the 21st century digital citizen process to be empowered, educated, and able to create almost everything.

2 Related Work

Research shows that digital fabrication technologies afford many opportunities for empowerment [3], development of 21st century skills [1] and even development of the self-perception as a digital citizen [22]. However, in education the potential to deploy these benefits for learning by creating concrete artefacts using digital fabrication seems to be lacking. Several reasons for this have been identified as it has happened when information and communication technologies are more or less successfully implemented in the classroom. For instance, challenges of technology availability, teachers' technological, pedagogical and content knowledge confidence of using the technologies, teachers' mindset, attitude as well as training, to name a few, are widely reported (see Häkkinen et al., [11], Mäkitalo-Siegl et al. [16], and Valtonen et al. [25, 26]). A big barrier appears when teachers have little knowledge about how to utilise a technology in a meaningful way in their educational activities, particularly when teachers have to change the pedagogy used in their classroom in order to utilise the technology [16]. Put differently, teachers hardly take in active use a new technology in their classrooms when there is no meaningful integration with the content students learn or pedagogy to use it [14]. It is also necessary to state that students need support and guidance when new or uncommon learning approaches are implemented in their classrooms and therefore, preparing teachers with clear instructions is one way to come up with good learning experiences and results [16]. On top of individual teacher's lack of skills or confidence, there are but only few networking opportunities where teachers can exchange experiences and materials when integrating new technologies in their activities [28]. Hence, in Europe, projects such as the eCraft2Learn[3] have explored the possibility to make knowledge generated by existing maker communities more accessible to the teacher community [29].

There have been several attempts to meaningfully introduce digital fabrication in educational contexts, nevertheless. For example, Blikstein and Krannich [4] report on attempts to introduce digital fabrication and making in formal and informal education arena from the perspectives of practical experiences as well as research approaches. In addition, Katterfeldt et al. [13] highlight their results of deploying constructivist learning environments for digital fabrication using programmable construction kits. They report on the use of digital fabrication technologies to foster deep and sustainable learning (*Bildung*). In a special issue on digital fabrication in education for the International Journal of Child-Computer Interaction, Iversen et al. [12] put forward contributions on bringing digital fabrication environments to education from several perspectives. These perspectives include the need of new curricula, social protocols, facilitation and tools demonstrating the scientific community interest to establish and reap the benefits of digital fabrication in the educational terrain.

Of particular interest to our work, Smith et al. [22, 23] propose a design thinking approach to digital fabrication, stressing the creative process from ideation to prototype. The work of Smith et al. [23] highlights three barriers to introducing digital fabrication in education from the perspective of the teacher through the design thinking approach lens: the complexity of the design process, the management of technologies and materials and the issue of balancing the different roles teachers need to assume during the design process. Nonetheless, design thinking is a concept familiar to designers and teachers who work in arts and crafts, as well as computer science. However, the problem occurs when design thinking is not well known among classroom and subject teachers who work in subjects such as Math, History, Biology, English or Social sciences. This is not to say that there is no design going on in today's classrooms. But that the benefit of design thinking structured approach is lost, if teachers cannot see a way to integrate design work with their work on subject domains.

[3] See H2020 eCraft2Learn project information at https://project.ecraft2learn.eu/who-is-it-for/.

Hence, we see that the challenges outlined in the contributions mentioned above are still to be fully met. The question of how do we reach the demands of 21st century work and digital society life within the educational arena to provide children with the necessary skills for *Bildung* and for becoming capable digital citizen is still a wide research field. Taking a pedagogical perspective stand, we propose a five-stage craft- and project-based approach to digital fabrication and making that includes elements of design thinking and inquiry-based learning to facilitate the inclusion of digital fabrication and making in formal and informal educational settings.

3 The Pedagogical Model

Our approach combines elements of design principles and inquiry-based learning to form methodological steps for the inclusion and utilization of technology for the creation of computer supporting artefacts. The aim is to produce a craft- and project-based pedagogy, which includes something familiar for teachers with some new elements for them to learn and implement in the classrooms. This pedagogy is derived from the inquiry-based learning methods and expanded to include both the knowledge and the specific product, which needs to be created through design thinking.

The inquiry-based learning starts by posing questions, problems or scenarios and a coach or facilitator of the learning experience usually supervises the process. Students then identify and research issues and questions to develop their knowledge or solutions. The process is intrinsically argumentative, where the students create questions, obtain supporting evidence to answer the questions, explain the evidence collected, connect the explanation to the knowledge obtained from the investigative process and finally create an argument and justification for the explanation. In order to enhance this process of inquiry, we draw from the design thinking method as hands-on counterpart.

We have seen design thinking used for digital fabrication deployment in education [22, 23]. Design thinking is a method for practical, creative solutions of problems with the intent of improve future results. Design thinking is a form of solution-based or solution-focused thinking that has been suggested for use in schools in a variety of curricular ways, as well as for redesigning student spaces and school systems [2]. Design thinking is employed to promote creative thinking, teamwork, and student responsibility for learning. Since the students work together in groups, design thinking in education also encourages team learning. Design thinking involves understanding with the learners exploring the topic; observing the environment; generating alternative point of view to enhance understanding; ideating solutions with an emphasis on creativity and enjoying the process; prototyping to investigate the ideas generated during ideation; and testing the ideas iteratively.

Our project deploys *a craft- and project-based* methodology, a combination of the inquiry-based learning with design thinking through the use of DIY and 3D printing technologies in order to enhance the pedagogical outcomes of the individually applied inquiry-based and design thinking methods. The craft- and project-based methodology involves the following processes (see Fig. 1):

Stage 1: Ideation Through World Exploration. In order to find out what kind of challenge they, the students, or someone else is facing in their daily life. The students could explore the world physical (e.g., taking pictures, exploring situations outside the classroom, etc.) or virtually (e.g., through online support community discussion) and then decide what their challenge will be. Alternatively, to facilitate the integration with school curricula, the teacher coach can pose an open problem taken from their subject matter for the students to come up with solutions to it. The teacher coach guides this process.

Stage 2: Planning. Once the challenge has been defined, the students collect information to make a project plan – they can get feedback from the teacher coach on their project plan and on the roles for group members. This they can also perform in parallel with ideation or creation, and be revised as often as needed according to new information obtained or ideas clarified as the project develops.

Stage 3: Creation. In this stage, the students embark on the co-design and co-creation of their computer-supported artefact solutions through the application of digital fabrication technologies. Visualisation and simulation of the designs are also important parts of this stage. The students can also utilise crafting and recycled materials (e.g., pet bottles, hot glue, cardboard, tape, etc.) alongside 3D printed parts.

Stage 4: Programming. Once the computer-supported artefacts are built, or as a parallel activity, the students define suitable scripts (high-level programming language) for the functioning of their artefacts. Software debugging and integrated SW/HW simulation are two other parts of this stage. We use Snap4Arduino programming environment in our piloting tests.

Stage 5: Sharing. In order to enhance motivation, the students are encouraged to share and showcase their projects and implementation ideas. This is done through the open (online) community or through peer sharing in the classroom. In return, the students have the opportunity to receive feedback from peers and as well as the community of practice worldwide (e.g., Arduino and Ultimaker).

The stages of the pedagogical model develop through collaborative interactions and co-creation, with every student bringing their strengths and interests to the realisation of the activity. The methodology also fosters students' self-regulation in terms of facilitating their understanding of what they already know and what they still need to know in order to bring their ideas and plan to life, as well as the management of time and resources. Self-reflection is also fostered in the methodology as students share their experiences with their peers, teachers and the open community – they need to understand what they have done and how in order to explain it to others.

Lammer et al. [15] also proposed a 5-step plan approach where children and researchers work together in dedicated workshops to support children's innovative and creative process with the aim to become robotic product designers. Our approach differs from Lammer et al.'s in the implementation and the aim of the expected outcome. In our methodology children immerse themselves in curricular tasks in their classrooms (or extra-curricular tasks in their after-school clubs) and utilise technology to propose solutions to those tasks. Children are not expected to become designers of

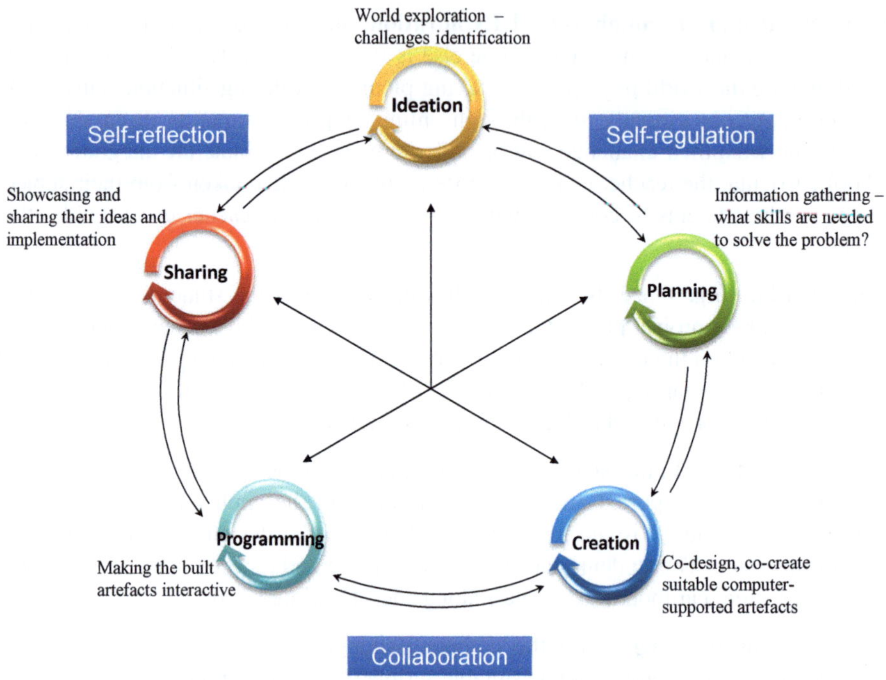

Fig. 1. Five-stage pedagogical approach of our project

robots but empowered *makers* of a variety of technological artefacts. Furthermore, out approach emphasises teamwork and collaboration during the entire process.

Furthermore, it is important to emphasise the intertwined nature of the pedagogical stages in the model proposed. That is, as a guideline the model can be used to facilitate the logical understanding of the activities that take place, however any stage can be carried out in combination with or as a parallel activity to any other stage. Therefore, their work does not have to be linear. Within project-based learning, students have the freedom to choose the subject matter and to define the central content of the project they want to work with. It is important that the projects can be linked to the authentic contexts of the real world. In *formal* education context the topic of the project could be taken from the subject curriculum (e.g. photosynthesis, security, geographical phenomena, etc.) or be an integration of several subject areas. In *informal* educational context, more freedom can be given to students to choose any topic or theme they feel curious about. In both application contexts, the resources available (e.g., digital components, recycled materials, etc.) need to also be taken into consideration.

4 Children's Creations

In our work, we aim at using craft- and project-based activities as part of the school curricula in formal settings as well as part of informal education activities (e.g. after-school clubs, Fablabs and maker spaces, etc.). This is so in order to foster creativity and the integration of arts with science, technology, engineering and math subjects within a sophisticated technical environment and using crafts materials (e.g., wood, pipe cleaners, construction paper, cotton balls, glitter and so forth). We recruited teacher volunteers from local schools and gave them 12 h of training on the pedagogical concepts of constructivism, constructionism, DIY philosophy, maker movement, design thinking and inquiry-based learning, which form the basis for our craft- and project-based methodology. The concepts of the transformation of the teacher's role to that of a coach or facilitator of the learning experience was also discussed during the training. Furthermore, they were introduced to digital fabrication technologies (e.g., 3D modelling with TinkerCAD, 3D printing, Arduino boards, resistors, light emitting diodes and other electronic components, and so forth) and to basic programming with Snap4Arduino programming environment.

We trialled our craft- and project-based pedagogical approach to digital fabrication and making through two curricular activities at a junior high school in Eastern Finland, with the assistance of the classroom teachers. The activities formed part of the pilots of our eCraft2Learn project. Three classes of school students participated in the activities voluntarily. One class was given the task to represent and model through technology the concept of photosynthesis, as part of their biology subject matter. The other two classes were tasked with the representation of the concept of security in society and how technology is applied, as part of their social studies subject matter. The pedagogical objectives of both tasks were therefore interdisciplinary, fostering students' understanding of biological processes and their social awareness respectively, as well as their digital literacy. Students were 14 years old in 8th grade (photosynthesis) and 15 years old in 9th grade (security). The biology class had 18 students (14 girls/4 boys), and in social studies one class had 13 students (all girls) and the other 15 (3 girls/12 boys) according to their study plans in the autumn term 2017. Students worked in their projects during school hours allocated to social studies subject matter (45 m/session, 20 sessions) and to biology subject matter (45 m/session, 7 sessions). They divided themselves into groups of 4 or 5 students each. They had no previous use of digital fabrication or making technologies in the classroom. The data was collected through photos, videos and recordings of the students' working sessions, semi-structured interviews with teachers and students about the activities, as well as the researchers' notes [13]. All ethical procedures for data collection were followed and teachers and students alike consented for their data to be collected and used for research purposes and for improving the learning environment and activities. Each class was introduced to the basic concepts of programming with blocks in Snap4Arduino environment as well as 3D modelling and printing in introductory hands-on lessons.

4.1 Ideation and Planning

At the start, teachers briefed the students about the background concepts of photosynthesis (i.e., including the basic structure and physiology of plants and organisms) and the social security issues that they would be investigating, in their respective classes. Students were then encouraged to develop a design incorporating those concepts using the technologies that they have been introduced to. As scaffolding, students were given written information on different electronic components that were available for them to use. Figure 2 shows students groups engaged in ideation and planning activities.

Fig. 2. (Left) a group's sketch of their photosynthesis model. (Right) a group brainstorming ideas and planning their way forward to model security in society

4.2 Creation and Programming

During the creation or making of their models, students employed a variety of materials including crafting materials (e.g., hot glue, ropes, etc.), recycled materials (e.g., cardboard and pet bottles) as well as 3D printed elements. The students also familiarised themselves with basic programming blocks, in Snap4Arduino programming environment enough for them to expand their knowledge while tinkering with their creations in order to make the artefacts interactive. For instance, based on the idea that photosynthesis only occurs when the right conditions of light, oxygen and water are present, one group representation of photosynthesis used sensors to indicate when the water levels of the plant were low (i.e., below a threshold) and flashed a visual alarm (using LEDs) to indicate that photosynthesis could not occur without water. In the social security classes, students became familiar with existing security systems in society by exploring how security is maintained for example in museums, banks, shops and houses. One student team was interested in exploring how security was managed in old castles and created their model of security from this inspiration. Figure 3 shows student groups in the process of making their models.

Fig. 3. Student groups working on the physical creation of their photosynthesis (top two rows) and security system models while making their artefacts interactive through programming

4.3 Sharing

At the end of the trials, a final questionnaire was distributed among the students of each class in order to survey their perceptions on their innovativeness, creativity and problem-solving skills development. The eCraft2Learn project was approved to carry out research with secondary school students and teachers by the Ethical Committee of the University of Eastern Finland. All participants in the study understood that their participation was voluntary and could freely withdraw at any time. Consent forms were obtained from the teachers, parents/guardians, and students prior the pilots. The data collected has been anonymised for research purposes.

In the biology class, 15 students returned the questionnaire. 13/15 students reported being innovative during the project, and 15/15 reported being creative. 12/15 reported feeling more comfortable than before to tackle problems. In the social security classes, 13 students returned the final questionnaire. All of the students reported being

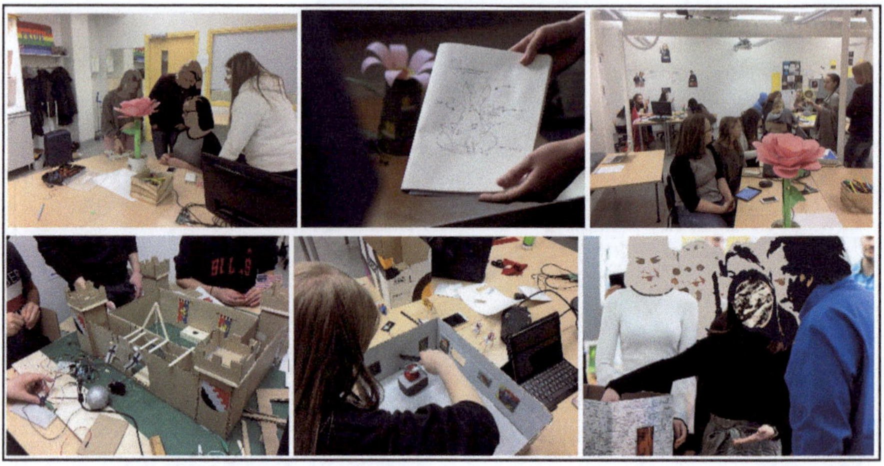

Fig. 4. Groups showcasing their model representation (photosynthesis in the top row).

innovative and creative during the project; whereas 12/13 students reported feeling more comfortable than before to tackle problems. This gives us an indication of the students' self-reflected views on their 21st century skills development through digital fabrication and making integrated in their normal curricular activities.

5 Teacher's Role

In our work, we consider the transformation of the teacher role to that of a coach or facilitator of the learning experience. To some extend this agrees with the notion of teachers as 'orchestrators' of the learning experience [5]. We believe, however, that this orchestration needs to be achieved in student-centred learning practices, where the student is the protagonist of the learning experience. The integration of digital fabrication in education brings with it new perspectives for students to learn. As such, personalised learning provides students the opportunity to learn in ways that is preferable for themselves by setting their learning objectives for their own learning. Each student may have different objectives, but the teacher should apply differenced strategies to promote the personal potentiality. Teacher's role is acting more as a coach. For students this means that self-regulation is a fundamental skill and not only the cognitive, but all dimensions of learning will be in focus (emotional, social, life experience etc.). (See Fullan [7] and Patrick [19]). Personalised learning sets challenges for a teacher. For instance, how to support deeper learning for each learner, adapted to each individual's knowledge level and expectations or how to connect learners' world and day-to-day experiences with in-class learning are still challenges teachers face [10].

Integrating digital fabrication in classroom activities could tackle such challenges, albeit perhaps creating issues of self-efficacy perception from the teacher's perspective (as reported by Smith et al. [23] for instance). However, these challenges are not trivial for teachers who expect themselves to be knowledgeable all the time. The sheer number of possible scenarios and questions makes it unlikely that teachers know it all. In such situations, being a coach can mean to help students finding the answer or parts thereof by directing them to external resources that could contain the required information. On the teachers' side, the skill of searching effectively is best acquired by being tinkerers themselves, realizing their own projects and developing a sense for when to 'move the goalpost' and when to stick to it because one is confident that there must be a solution. This level of self-efficacy can only be acquired through experience. As noted earlier, in teacher's professional mindset, the beliefs about learning and knowledge might be different compared to the approach used in the eCraft2Learn and therefore, building the visible learning process with detailed instructions in different stages for enhancing teachers to implement these ideas in the classrooms. Our eCraft2learn project supports teachers in gaining this experience trough offering training and space to explore the benefits of digital fabrication technologies alongside the pedagogical model described before. The pedagogical model was implemented and tested during two pilots that deployed curricular tasks like the ones shown in previous sections, as well as activities in after-school clubs. From the pilots we could obtain feedback from nine teachers who participated. Comments and feedback were collected through semi-structured interviews as well as researchers' notes during the training sessions, documented by observers and moderators of the sessions. Following a few excerpts of the teachers' views:

- Teachers agreed that *learning-by-making* supports learning. When students notice that something does not work or a different solution could have been better, they will analyse the problem, search for information and the next time they manage to do things more smoothly.
- Teachers admitted that they have quite limited knowledge and a narrow or stereotypical view of technology. One teacher was amazed how many new possibilities for the creation of technology he d iscovered during the project. This related primarily to the fact that sensors provided endless opportunities to children who wanted to modify the behaviour of a 'robot', that could take the shape of a flower.
- Teachers also found that some fundamental dependencies had to be acknowledged, such as knowing basic programming constructs and the use of variables. Otherwise introducing those concepts would be quite time consuming. This does not mean that all programming needs to be taught prior to a making project, and an overview session was considered to be sufficient.
- Teachers also noticed how challenging the ideation phase could be, since they had limited background information about the workings of different sensors and other electronic components. This highlights the need to match the novelty of a topic with the appropriate learning strategy. For example, it is known that for novice learners worked-examples can be very effective [20]. Over time, with more contextual knowledge available, the need for examples decreases and problem-based approaches become more effective. From the teachers' point of view, there had to be a balance between student- and teacher-led learning. When introducing difficult content, a teacher cannot shift all the responsibility onto students. It is necessary to use a degree of scaffolding depending on students' current level of understanding.

- Finally, teachers noticed the importance of time-management. Students should set goals consciously and be reminded to monitor their own work in relation to the goals. Too tight schedules can excessively limit the work and learning about effective time-management practices.

6 Concluding Remarks

Here we have presented our approach to bring digital fabrication to educational context. Our approach puts forward a craft- and project-based pedagogy in order to facilitate the intake of digital technologies into formal school grounds. We presented examples of the students' steps and works through the five-stage pedagogy from the ideation, planning, creation and programming to the sharing of their artefacts. We could observe the integration of digital fabrication in school curriculum activities through the application of our pedagogical approach. Teachers' role as a coach that orchestrate the learning experience during the process is emphasised in this approach. We highlighted the importance of working under a pedagogical umbrella that teachers can assimilate and understand in order to facilitate the inclusion of digital fabrication in the classroom. We highlight that compared to other approaches and projects that have attempted to deploy similar ideas of constructivism and constructionism in education in the past 30 years (see for instance a report on Project UPDATE [24]), the success has been but limited. We could argue that changing only the pedagogical approach is not sufficient to make the transformation of learning and teaching via hands-on learning methods. In addition to it, teachers need to change their beliefs about learning and knowledge, fundamentally. Through the eCraft2Learn project, we have aimed at fostering an attitude change towards technology integration in education taking a broad view of technology making and application to non-technical subjects such as history and biology. The results are encouraging.

Acknowledgements. The work presented here was carried out under the H2020 eCraft2Learn Project. This project has received funding from the European Research Council (ERC) under the European Union's Horizon 2020 research and innovation programme, grant agreement No. 731345.

References

1. Alimisis, D.: Educational robotics: open questions and new challenges. Themes Sci. Technol. Educ. **6**(1), 63–71 (2013)
2. Bekker, T., Bakker, S., Douma, I., Van Der Poel, J., Scheltenaar, K.: Teaching children digital literacy through design-based learning with digital toolkits in schools. Int. J. Child Comput. Interact. **5**, 29–38 (2015)
3. Blikstein, P.: Digital fabrication and 'making' in education: the democratization of invention. In: FabLabs: of Machines, Makers and Inventors, vol. 4, pp. 1–21 (2013)

4. Blikstein, P., Krannich, D.: The makers' movement and FabLabs in education: experiences, technologies, and research. In: Proceedings of the 12th International Conference on Interaction Design and Children, pp. 613–616. ACM, June 2013
5. Dillenbourg, P., Jermann, P.: Technology for classroom orchestration. In: Khine, M., Saleh, I. (eds.) New science of learning, pp. 525–552. Springer, New York (2010)
6. Eisenberg, M.: 3D printing for children: what to build next? Int. J. Child Comput. Interact. **1**(1), 7–13 (2013)
7. Fullan, M.: Personalized learning (2009). https://michaelfullan.ca/wp-content/uploads/2016/06/Untitled_Document_16.pdf
8. Gershenfeld, N.: How to make almost anything: the digital fabrication revolution. Foreign Aff. **91**, 43 (2012)
9. Gutek, G.L.: The Montessori method: the origins of an educational innovation: including an abridged and annotated edition of Maria Montessori's The Montessori method. Rowman & Littlefield Publishers (2004)
10. Hargreaves, A.P., Shirley, D.L. (eds.): The Fourth Way: The Inspiring Future for Educational Change. Corwin Press, Thousand Oaks (2009)
11. Häkkinen, P., Järvelä, S., Mäkitalo-Siegl, K., Ahonen, A.K., Näykki, P., Valtonen, T.: Preparing teacher students for twenty-first-century learning practices: a framework for enhancing collaborative problem-solving and strategic learning skills. Teachers Teach. Theory Pract. **23**(1), 25–41 (2016)
12. Iversen, O.S., Smith, R.C., Blikstein, P., Katterfeldt, E.-S., Read, J.C.: Digital fabrication in education: expanding the research towards design and reflective practices. Int. J. Child Comput. Interact. (2016). https://doi.org/10.1016/j.ijcci.2016.01.001
13. Katterfeldt, E.S., Dittert, N., Schelhowe, H.: Designing digital fabrication learning environments for Bildung: implications from ten years of physical computing workshops. Int. J. Child Comput. Interact. **5**, 3–10 (2015)
14. Koehler, M.L., Mishra, P., Cain, W.: What is technological pedagogical content (TPACK)? J. Educ. **193**(3), 13–19 (2013)
15. Lammer, L., Weiss, A., Vincze, M.: The 5-step plan. In: Abascal, J., Barbosa, S., Fetter, M., Gross, T., Palanque, P., Winckler, M. (eds.) Human-Computer Interaction – INTERACT 2015. INTERACT 2015. Lecture Notes in Computer Science, vol. 9297. Springer, Cham (2015)
16. Mäkitalo-Siegl, K., Kohnle, C., Fischer, F.: Computer-supported collaborative inquiry learning and classroom scripts: effects on help-seeking processes and learning outcomes. Learn. Instruct. **21**(2), 257–266 (2011)
17. Nunez, J.G.: Prefab the FabLab: rethinking the habitability of a fabrication lab by including fixture-based components. Doctoral dissertation, Massachusetts Institute of Technology (2010)
18. Papert, S.: Mindstorms: Children, Computers, and Powerful Ideas. Basic Books Inc., New York (1980)
19. Patrick, H.: Social self-regulation: exploring the relations between children's social relationships, academic self-regulation, and school performance. Educ. Psychol. **32**(4), 209–220 (1997)
20. Renkl, A.: The worked-out examples principle in multimedia learning. In: Mayer, R.E. (ed.) The Cambridge Handbook of Multimedia Learning. Cambridge University Press, Cambridge (2005)
21. Schelhowe, H.: Interaktionsdesign für reflexive Erfahrung. Be-greifbare Interaktionen, pp. 253–272. Transcript Verlag, Bielefeld, Germany (2012)
22. Smith, R.C., Iversen, O.S., Hjorth, M.: Design thinking for digital fabrication in education. Int. J. Child Comput. Interact. **5**, 20–28 (2015)

23. Smith, R.C., Iversen, O.S., Veerasawmy, R.: Impediments to digital fabrication in education: a study of teachers' role in digital fabrication. Int. J. Digital Literacy Digital Competence **7**(1), 33–49 (2016)
24. Todd, R.: A new paradigm for pre-engineering education: design and technology education K-12. In: Proceedings of IEEE Frontiers in Education Conference - FIE 1993 (1993). https://doi.org/10.1109/fie.1993.405501
25. Valtonen, T., Kukkonen, J., Kontkanen, S., Mäkitalo-Siegl, K., Sointu, E.: Differences in pre-service teachers' knowledge and readiness to use ICT in education. J. Comput. Assist. Learn. 1–9 (2018). https://doi.org/10.1111/jcal.12225
26. Valtonen, T., Mäkitalo-Siegl, K., Kontkanen, S., Pöntinen, S., Vartiainen, H.: Facing challenges with new teachers' use of ICT in teaching and learning. Bull. IEEE Tech. Committee Learn. Technol. **14**(4), 46–49 (2012)
27. Voigt, C., Montero, C.S., Menichinelli, M.: An empirically informed taxonomy for the maker movement. In: Bagnoli, F., et al. (eds.) International Conference on Internet Science, pp. 189–204. Springer, Cham (2016)
28. Voigt, C.; Schön, S., Hofer, M.: Innovation management in schools: barriers and enablers to making as educative practice. In: Multikonferenz Wirtschaftsinformatik (MKWI) (2018)
29. Voigt, C., Mair, S., Unterfrauner, E.: Hacking the knowledge of maker communities in support of 21st century education. In: International Conference on Internet Science, St Petersburg. Springer (2018)
30. Wikipedia contributors: Maker culture. In: Wikipedia, The Free Encyclopedia, 17 July 2018. https://en.wikipedia.org/w/index.php?title=Maker_culture&oldid=850779485. Accessed 31 July 2018

Badges Are Back! - Fostering Self-assessment During Personalised Learning in Making and Digital Fabrication

Francesca Agatolio[1], Calkin Suero Montero[2(✉)], Michele Moro[3], Andrea Cavazzini[3], and Emanuele Menegatti[3]

[1] Department of Psychology, University of Padova, Padua, Italy
francesca.agatolio@studenti.unipd.it

[2] School of Educational Sciences and Psychology, University of Eastern Finland, Joensuu, Finland
calkins@uef.fi

[3] Department of Information Engineering, University of Padova, Padua, Italy
{mike, emg}@dei.unipd.it,
andrea.cavazzini@studenti.unipd.it

Abstract. Badges have been used in education and leisure activities to accredit knowledge or learning for a long time. Digital badges in particular have been reported to become a gamification element fostering students' engagement and motivation. In our work, we explore the use of digital badges during digital fabrication and making activities as a way to foster personalised learning paths within the eCraft2Learn learning ecosystem. In this paper, we describe the development of the badge system implemented in the eCraft2Learn digital platform. The system supports students' self-evaluation and promotes a communicative process between students and teachers for badge allocation. Visual representations of the badge system are provided here alongside the underlying pedagogical justifications behind the system development.

Keywords: Technology innovation · Badges in education · Digital fabrication and making

1 Introduction

Personalisation is a broader concept than individualization or differentiation – personalised learning means that the learning process is driven by the learners' interests, needs and timeline. This means also adjusting the pace and the approach taking into account the learners' specificities. Thanks to new digital technologies, learning can be done anytime, anywhere or any place. In personalised learning, learners' previous knowledge, competence, life and work skills both in formal and informal settings should be considered. Within this frame, learners do have the opportunity to learn in ways that are preferable for themselves and also set their learning objectives for their own learning. In this sense, in personalised learning the teacher is acting more as a coach, rather than an instructor. This means that learner's self-regulation is a

M. Moro et al. (Eds.): Edurobotics 2018, AISC 946, pp. 83–91, 2020.
https://doi.org/10.1007/978-3-030-18141-3_7

fundamental skill and not only the cognitive dimension, but all dimensions of the learner should be in focus (emotional, social, life experience etc.) (See [8, 18]).

Miliband [14] has put forward five key elements that should be brought with personalised learning:

- knowing the strengths and weaknesses of individual students,
- develop competences and confidence in the learner through teaching and learning strategies based on individual needs
- create curriculum choice that engages and respects students
- put forward a radical approach to school organization
- get the community, local institutions and social services

We propose an approach to foster personalised learning through the use of badges during making and digital fabrication activities in order to provide a suitable solution to the first two challenges listed above. With our approach we do not only aim at assisting the students understand their strengths and weaknesses but also assisting them to develop their confidence during the learning experience. Furthermore, we also aim at fostering in students the capability to actively reflect on their learning process. This cannot be achieved if badges are used in the usual way, i.e. the teacher directly assigns the badges to the students on the basis of the outcomes of their activities and results during the implementation of their projects. Going beyond this, we propose in this paper to use badges within a self-assessment framework, where the students are requested to provide the evaluation of their own skills as they have developed during the practical activities. The self-assessment is composed of a score (in a scale from 1 to 5) and an textual justification for the score. The teacher reviews this self-evaluation and if necessary revise it, explaining to the students the reasons for the modifications. The teacher evaluations are converted into graphical scores in the form of badges.

2 Related Work

Personalised learning has been fostered by a number of pedagogical approaches as well as tool. For instance, McLoughlin and Lee [12] discuss the feasibility of applying the social web tools (web 2.0) to foster personalised learning and self-regulation, through social media tools, such as Twitter™, urging for a pedagogical change to facilitate the integration of such social software tools. Peirce et al. [19] propose an adaptive educational game architecture to foster personalised learning through gaming, avoiding the one-size-fits-all approach present in many educational games. Within the gaming approach, digital badges have also been introduced as a tool for boosting student's motivation, providing recognition of one's credentials as well as evidencing achievement [9]. With this in focus, we propose a badge system that is based on the students' self-assessment. Our system is developed as part of the eCraft2Learn learning ecosystem, as a tool for fostering personalised learning.

3 Badge System for Personalised Learning

The eCraft2Learn project (https://project.ecraft2learn.eu/) is a H2020 European project dedicated to develop an innovative learning ecosystem able to integrate making and digital fabrication, the essence of the recent maker movement, in a pedagogically sustained framework. The project's target are 13–17 years old students. The project can be summarised by mentioning the following main aspects [1]:

- a project-based oriented pedagogy able to foster engagement, motivation, self-awareness, team-working, both abstract and practical abilities, in students, with a specific attention to STEAM teaching-learning purposes;
- a Unified User Interface (UUI) integrating open source or at least free software capable to support all the phases of development from ideation to evaluation and sharing of the produced artefacts;
- a hardware architecture based on a low-cost, fully featuring board, like Raspberry PI3, used as developing system and running the UUI, and a do-it-yourself electronic board like Arduino to connect small components like motors, lights and sensors;
- a 3D printing service to provide 3D printing capabilities able to enrich the production of artefacts with self-designed parts and to empower creativity;
- a rich repository of open educational resources to support teachers and students, both at a personal and team level, in using the provided tools, promoting an autonomous learning progression and an easy way to overcome basic technical problems;
- a series of use cases to be used to get started with the system, to show how to fully exploit its educational potential.

The learning path is partly guided by a five stages pedagogical approach that we have put forward during the project design (ideation, planning, creation, programming, sharing; see Fig. 1). These stages do not form a linear sequence of steps which have to be strictly and orderly followed by the learners, but learners can use them as a frame for structuring their work, adapting their path according to what they need or they want a deeper understanding about during the development of a project. This is a first form of personalisation that the learning ecosystem allows and promotes.

In spite of its relative simplicity, an Arduino board equipped with actuators and sensors can provide the basis for meaningful robotic experiences: in fact, students must refer to physical phenomena and uncertainties, which are present in a real environment in contrast with the usual ideal essence typical of virtual environments. This means that they have to sense some physical aspect around their artefact and program accordingly its actions, something that solicits to have a better understanding of what is about, to refine solutions during the development to obtain the expected behaviours and to reflect on the reached solutions at personal level and within the team.

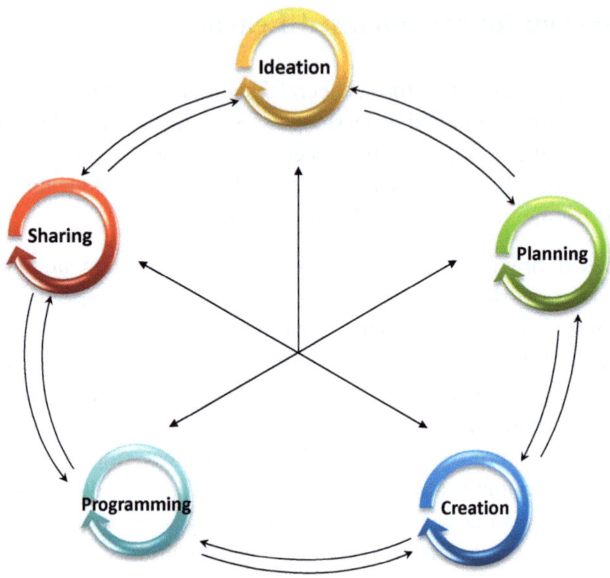

Fig. 1. The five stages of the eCraft2Learn methodology (adapted from Suero Montero et al. [21])

During small-scale pilot testing planned within the project, we observed that some teams decided to be challenged in building a real mobile robot. In this case, for time restrictions, they were provided with a simple Arduino-based robot kit that may include motors, LEDs, distance and light sensors, but even in this situation, they had the opportunity to personalise their artefact with some 3D-printed extension. In the end, we think that the ecraft2Learn ecosystem may represent an effective synthesis between educational robotics and the maker movement, providing a full spectrum of state-of-the-art tools and a meaningful teaching-learning methodology.

As briefly introduced above, the entire eCraft2Learn ecosystem has been developed with the aim of making students the protagonists of the learning process, taking into account their interests and attitudes; for this purpose, it focuses on the concept of *personalised learning*. It is important to notice that the learning ecosystem has been designed to encourage collaborative group work in order to foster the multidisciplinary development of tasks as well as 21st century skills such as collaboration, problem solving and creativity. In order to actively observe their own learning progression and, at the same time, to receive a sort of immediate recognition and gratification of their efforts, the learning ecosystem offers integrated evaluation and badge system based on students' self-assessment. The design of the badge system followed the same principle of personalisation aiming at being intuitive to use. In our work, we use the self-assessment definition as "… a wide variety of mechanisms and techniques through which the students describe (i.e., assess) and possibly assign merit or worth to (i.e., evaluate) the qualities of their own learning processes and products" [16]. Self-assessment is concerned with the ability to correctly evaluate one's own work, acknowledging both strengths and weaknesses (an important aspect of personalised learning); as such, it is a key ability that students should acquire. In fact, the numerous studies on the impact of the self-assessment on learning show that the practice has positive effects both on results and on self-esteem [2, 13, 15]; moreover, they show a positive correlation with the ability of self-regulated learning [3]. However, Panadero et al. [17] underline that the effectiveness of the self-assessment practice depends on how it is taught; it is not sufficient to ask the students to give a "grade" to their work, but it is fundamental to make them develop an awareness about the critical issues of subjects and about the chance for further improvement.

3.1 Self-assessment

The badge system contains a self-assessment mechanism through which the students can evaluate their progress. Taking into consideration the suggestions of key teacher informants, we identified six categories for self-assessment, corresponding to the various aspects of a digital fabrication project development: *building electronic circuits*, *3D design*, *programming*, *presentation of the project outcomes*, *quality of the teamwork* and *originality*. Parameters provided for each of these categories guide the students to achieve a reasoned and objective evaluation of their work (Fig. 2).

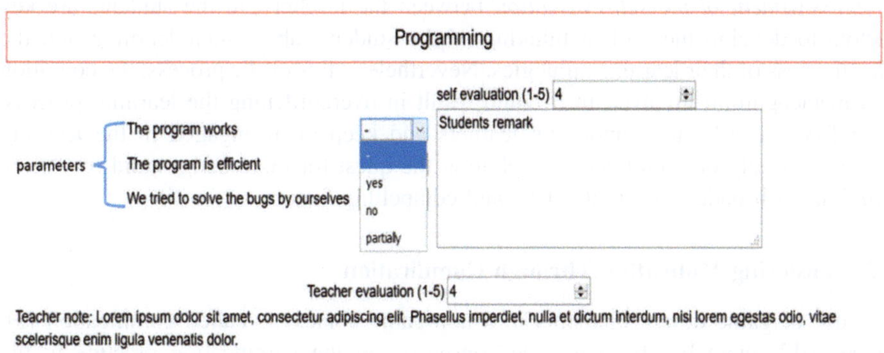

Fig. 2. The self-assessment mechanism showing the parameters for the programming category evaluation

The system provides default parameters to use for self-evaluation; however, the teachers are able to provide their own parameters as well. Ideally, students and teachers should decide and set together the parameters, thus sharing the responsibility to represent the most relevant and meaningful aspects of the learning process. When asking the key teacher informants their opinion regarding the default parameters that define each category, we understood that flexibility was needed in terms of allowing the teachers to modify the categories and their parameters in order to tune the evaluation to the aspects that they considered relevant to be self-reflected and self-evaluated by the students.

The self-assessment system requires that the students first think about which parameters have been achieved; then, they are asked to assign to each of the six categories a score between 1 and 5 alongside a brief justification for their score. In addition, after each activity, the students are asked what they learnt new, and what they found too difficult or they didn't completely understand. For this reason, even if reference parameters are available, the score that the students assign to their work is not necessarily tied to their actual achievements. That is, the artefact created by the students in their project might not work perfectly, but the student could anyway self-assess a good mark, because they have learned other important skills; what is important is that students provide justification to their considerations. For this purpose, the interface has text boxes where students should summarise the rationale behind the numerical scores that they assign to their work.

The information obtained from these rubrics is precious to the teacher: on the one hand it delivers data on the efficiency of the self-assessment process, and on the other hand it allows the teacher to judge if the learning activity was properly calibrated, offering a comprehensive vision of the strong and weak points of their students. The teacher can confirm the student produced score or assign a different one in the self-evaluation interface, provided that the teacher justifies the change. The teacher modifications and justifications are visible to the students. This transparent evaluation from the teacher and this cyclic process of *self-evaluation* ⟷ *evaluation* ⟷ *assessment* ⟷ *assessment of the self-evaluation* between the teacher and the students are key factors to develop the critical thinking of the students about their learning and the effectiveness of their learning strategies. Nevertheless, this cyclic process, the cognition and metacognition involved in it could result in overburdening the learning process. Therefore, in order to support the learners and keep them engaged in the learning process, we rely on gamification exploiting the quest for more badges and for "*more stars*" in each badge as a motivating and compelling feature.

3.2 Fostering Motivation Through Gamification

The use of game design elements in a non-game context is called gamification [5]. Since 2013, there has been a wide consensus on the gamification practice in the educational environment, especially at the higher degrees of education: badges are an example of game mechanics, along with points, levels, progress bars, leaderboards, virtual currency, and avatars. As several studies report, the use of these mechanics have a great potential in increasing students' engagement and in motivating them to learn and train new skills [4, 6]. In particular, badges are proved to be effective in triggering competitive motivation [20], in improving learners' participation [7] and in enhancing learning, time management and carefulness [10].

In the eCraft2Learn digital platform of the learning ecosystem, the self-evaluation interface that we are presenting in this paper as a gamification process has been embodied into a set of badges. Indeed, the scores assigned by the teacher are elaborated by the eCraft2Learn system and converted into badges; they can also refer to the learning analytics system [22] in order to provide a more objective evaluation. Badges give students an immediate and easy to understand feedback about their progress in each category. The basic idea is outlined in Fig. 3. This graphical interface at the moment serves only as a working prototype. The badge interface will be (re-)designed by professional graphical experts in order to be in line with the rest of the graphical appearance of the eCraft2Learn digital platform and to be highly appealing and attractive for the students.

The badge system we designed has a twofold goal. On the one hand, it shifts the attention of the students from the practical activities they have to carry on to complete the project assigned to them, to the skills they are supposed to learn during this activity. This is achieved by the careful choice of the six categories for self-assessment, which were defined with the help of key teacher informants involved in the eCraft2Learn project. Each of these six categories visually prompt the students to actively reflect on the fundamental skill they are supposed to assimilate abstracting their thinking from the technical/practical problems at hand. As one can notice, the six categories are divided into technical skills specific to digital fabrication and making activities (such as circuit building or 3D design) and the cross-skills to be acquired (such as presentation capabilities and quality of the teamwork).

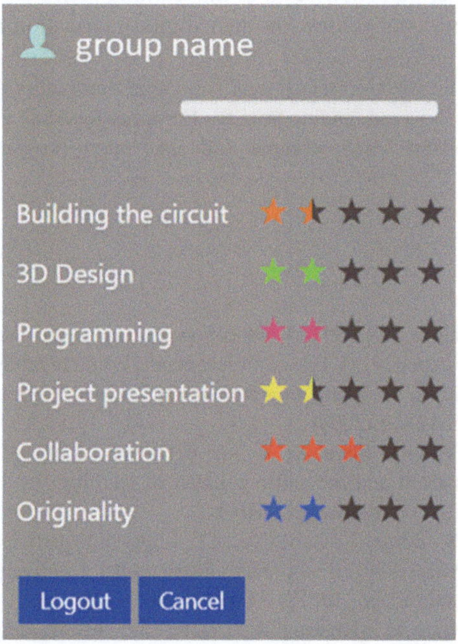

Fig. 3. The scores confirmed by the teacher are converted into graphical badges

On the other hand, the badge system motivates the students to complete their digital fabrication projects and the practical activities associated with the projects[1]. The more activities the students complete, and more carefully they implement them, the more

[1] For examples of these projects and activities please visit the eCraft2Learn website https://project.ecraft2learn.eu/. See also to the overview reported by Kahn et al. [11].

badges they will earn. Moreover, in order to earn more badges, they have to finish their work and critically revise and evaluate it, speculating on the achievements of their own learning. This process aims at fostering a reasoning on the outcomes of their learning. Additionally, the students are encouraged to develop (and express) a fair self-evaluation, because the teacher is evaluating both the technical achievements and the self-evaluation capabilities of the students.

4 Conclusions

In this paper we presented a badge system with an underline gamification aspect. The badge system is based on a student's self-evaluation mechanism in order to create a cyclic evaluation process between students and teachers. The proposed badge system with its corresponding self-assessment mechanism focuses on supporting students' self-evaluation process (i.e. to award a grade to one's own work), as well as fostering personalised learning through self-reflection, supporting also self-regulation. The system is expected to foster personalisation of the learning path while assisting students to develop motivation and engagement on their digital fabrication activities.

Acknowledgements. The work presented here was carried out under the H2020 eCraft2Learn Project. This project has received funding from the European Research Council (ERC) under the European Union's Horizon 2020 research and innovation programme, grant agreement No. 731345.

References

1. Alimisis, D., Loukatos, D.: STEM education post-graduate students' training in the eCraft2Learn ecosystem (2018). https://project.ecraft2learn.eu/publications/
2. Andrade, H., Valtcheva, A.: Promoting learning and achievement through self-assessment. Theory Pract. **48**(1), 12–19 (2009)
3. Andrade, H., Cizek, G.J.: Students as the definitive source of formative assessment: academic self-assessment and the self-regulation of learning. In: Handbook of Formative Assessment, pp. 102–117. Routledge (2010)
4. Barata, G., Gama, S., Jorge, J., Goncalves, D.: Improving participation and learning with gamification. In: Proceedings of the First International Conference on gameful design, research, and applications, pp. 10–17. ACM (2013)
5. Deterding, S., Dixon, D., Khaled, R., Nacke, L.: From game design elements to gamefulness: defining gamification. In: Proceedings of the 15th International Academic MindTrek Conference: Envisioning Future Media Environments, pp. 9–15. ACM (2011)
6. Dicheva, D., Dichev, C., Agre, G., Angelova, G.: Gamification in education: a systematic mapping study. J. Educ. Technol. Soc. **18**(3), 75–88 (2015)
7. Dominguez, A., Saenz-De-Navarrete, J., De-Marcos, L., Fernandez-Sanz, L., PageS, C., Martinez-Herralz, J.J.: Gamifying learning experiences: practical implications and outcomes. Comput. Educ. **63**, 380–392 (2013)
8. Fullan, M.: Personalized learning (2009). https://michaelfullan.ca/wp-content/uploads/2016/06/Untitled_Document_16.pdf

9. Gibson, D., Ostashewski, N., Flintoff, K., Grant, S., Knight, E.: Digital badges in education. Educ. Inf. Technol. **20**(2), 403–410 (2015)
10. Hakulinen, L., Auvinen, T.: The effect of gamification on students with different achievement goal orientations. In: 2014 International Conference on Teaching and Learning in Computing and Engineering (LaTiCE), pp. 9–16. IEEE (2014)
11. Kahn, K., Suero Montero, C., Voigt, C.: STEAM learning in formal and informal settings via craft and maker projects. In: Proceedings of the 17th ACM Conference on Interaction Design and Children, pp. 728–733. ACM (2018)
12. McLoughlin, C., Lee, M.J.W.: Personalised and self regulated learning in the Web 2.0 era: international exemplars of innovative pedagogy using social software. Australas. J. Educ. Technol. **26**(1), 28–43 (2010)
13. McMillan, J.H., Hearn, J.: Student self-assessment: the key to stronger student motivation and higher achievement. Educ. Horiz. **87**(1), 40–49 (2008)
14. Miliband, D.: Choice and voice in personalised learning. In: Schooling for Tomorrow Personalising Education, Chapter 1, Organisation for Economic Co-operation and Development (OECD) (2016)
15. Panadero, E., Alonso-Tapia, J.: Self-assessment: theoretical and practical connotations, when it happens, how is it acquired and what to do to develop it in our students (2013)
16. Panadero, E., Brown, G.T., Strijbos, J.W.: The future of student self-assessment: a review of known unknowns and potential directions. Educ. Psychol. Rev. **28**(4), 803–830 (2016)
17. Panadero, E., Jonsson, A., Botella, J.: Effects of self-assessment on self-regulated learning and self-efficacy: four meta-analyses. Educ. Res. Rev. **22**, 74–98 (2017)
18. Patrick, H.: Social self-regulation: exploring the relations between children's social relationships, academic self-regulation, and school performance. Educ. Psychol. **32**(4), 209–220 (1997)
19. Peirce, N., Conlan, O., Wade, V.: Adaptive educational games: providing non-invasive personalised learning experiences. In: 2008 Second IEEE International Conference on Digital Games and Intelligent Toys Based Education, pp. 28–35. IEEE (2008)
20. Pirker, J., Riffnaller-Schiefer, M., Gütl, C.: Motivational active learning: engaging university students in computer science education. In: Proceedings of the 2014 Conference on Innovation & Technology in Computer Science Education, pp. 297–302. ACM (2014)
21. Suero Montero, C., Voigt, C., Mäkitalo, K.: From digital fabrication to meaningful creations: pedagogical perspectives. Springer (to appear, 2019)
22. Toivonen, T., Jormanainen, I., Suero Montero, C., Alessandrini, A.: Innovative maker movement platform for K-12 education as a smart learning environment. In: Challenges and Solutions in Smart Learning, pp. 61–66. Springer, Singapore (2018)

Programming, Games and Social Robots

Exploring Escape Games as a Teaching Tool in Educational Robotics

Christian Giang[1,2(✉)], Morgane Chevalier[1,3], Lucio Negrini[2], Ran Peleg[4], Evgeniia Bonnet[1], Alberto Piatti[2], and Francesco Mondada[1]

[1] Mobots group of the Biorobotics Laboratory, Ecole Polytechnique Fédérale de Lausanne (EPFL), Lausanne, Switzerland
christian.giang@epfl.ch
[2] Department of Education and Learning (DFA), University of Applied, Sciences and Arts of Southern Switzerland (SUPSI), Locarno, Switzerland
[3] Haute École Pédagogique (HEP) du Canton de Vaud, Lausanne, Switzerland
[4] EdQuest, London, UK

Abstract. In the past years, educational escape games have raised interest in researchers and educators as a new game-based learning approach to break out from the traditional classroom routine. Recent work has demonstrated that educational escape games elicit high motivation and engagement on the part of the participants. Moreover, it has been shown that many participants experienced flow, a state of mind, which has been considered as beneficial for successful learning. However, due to its novelty, studies on the educational significance of such activities still appear to be sparse. In this work, we investigated whether escape games can be used as a teaching tool in the context of educational robotics, since both can be situated in the pedagogical currents of social-constructivism. To this end, we developed a prototype of an escape game using the educational robot Thymio and the visual programming language VPL and tested it with 61 subjects. Both quantitative and qualitative results showed that most participants highly appreciated the activity and agreed on the game's usability for teaching. Moreover, a great majority agreed on having experienced flow while playing. We, therefore, suggest that escape games provide a favorable framework for educational robotic activities, promoting particularly self-regulated and collaborative learning.

Keywords: Constructivism · Educational robotics · Game-based learning

1 Introduction

1.1 Background

Entry into and adherence to learning has always been a recurrent theme in education. In this context, learning motivation has been widely discussed [1, 2] and frameworks to promote engagement and motivation have been proposed [3]. In this context, game-based learning (GBL) approaches, capitalizing on the engaging character of games, have become more and more popular. In the past decades, particular attention has been

M. Moro et al. (Eds.): Edurobotics 2018, AISC 946, pp. 95–106, 2020.
https://doi.org/10.1007/978-3-030-18141-3_8

devoted to digital GBL, a field popularized by Prensky [4], aiming at integrating educational elements into computer games and/or digital simulations. Researchers and educators have agreed on the potential of such solutions, used for both instructional teaching [5] as well as for the exploration of new topics [6, 7]. However, GBL also includes non-digital examples, such as educational card, board or role-playing games, allowing to implement more tangible and human-centered activities. In recent years, escape games have become increasingly popular [8] and there have been attempts to exploit this concept for educational purposes [9]. In escape games, players usually team up to discover clues helping them to solve a series of tasks and puzzles. The game is typically (but not necessarily) played in a room locked from the outside, and the main objective is to find a way to escape from it by solving the presented puzzles (i.e., why they are often also referred to as escape rooms). Often, the game is embedded in a narrative and the tasks have to be solved within a given time limit (typically between 30–60 min). Besides the captivating character of escape games, which puts the players into an immersive problem-solving situation, the need for collaboration and communication to solve the puzzles as a team has been considered as a main appeal [9]. From a pedagogical point of view, escape games are a methodology based on a social-constructivist approach [10]: the learner is called to face new, rather difficult problems, which can be solved thanks to the interaction with peers and the support of the teacher. The role of the teacher in this approach is to structure the learning environment and provide instructional scaffolding to the learners, "*facilitating the students' interaction with the material and with each other*", as described by King [11]. Interestingly, educational robotics (ER) activities are often based on the same pedagogical concepts. Indeed, Papert's constructionism [12] can be considered as an application of Piaget's constructivism [13] to ER. Moreover, very often ER activities are proposed to small groups of students, prompted to collaborate and communicate to solve a given problem situation. We, therefore, argue that due to the pedagogical affinity of escape games and ER, the adoption of the former concept in the latter field could generate interesting and favorable learning situations. Indeed, there have been a few approaches to use educational escape games as a mean of instruction - however, none of them in the context of ER: while Vörös and Sárközi [14] developed an escape game to teach the physics of fluids to a group of gifted students, Kinio et al. [15] devised an escape game for the surgical education of medical students. Both works reported high engagement and enjoyment on the part of the participants and suggested escape games to be a promising approach to break out from the traditional classroom routine. Similar results were found by Hou and Chou [16], who developed a digital escape game (i.e., it was played on a computer) to teach the basics of electromagnetism to high school students. In addition to high engagement, the authors reported that participants experienced flow - a state of mind, which has been considered as beneficial for successful learning [17]. Despite those recent advances, scientific research on educational escape games is still in its infancy and literature about their efficacy and usefulness as an educational tool still appears to be sparse. To this end, we here present the prototype of an escape game using the educational robot Thymio and the visual programming language VPL. Our study aims at emphasizing how escape games can be effectively used as a teaching tool, particularly in the context of ER.

1.2 Research Questions

The goal of this study was to address research questions with focus on two aspects:

Players' Perception of the Game: How do individuals perceive escape games in the context of educational robotics? Do they experience flow during the activity? Are there differences between genders or between levels of experience?

Educational Potentials of Escape Games: How can escape games be used as a teaching tool for ER?

2 The Thymio Escape Game

For this study, we designed a prototype of an escape game using the educational robot Thymio [18]. The development was based on design guidelines proposed in previous works about learning games [19] and educational escape games [8]. The main learning objective was to introduce the players to the Thymio robot and the visual programming language VPL [20]. Participants played in groups of 3–5 people and had 30 min to solve the quest. The game started by bringing the group into a dark room and showing them a video to introduce the scenario: the local energy supply has been attacked and the task for the group was to reactivate the main energy source, a battery powered by mobile robots. In order to accomplish this task, the players had to find three Thymio robots hidden in the room, each characterized by a different color (yellow, blue and red). After the robots were found, they had to be programmed using VPL. The goal was to make them move inside the three compartments of the battery, where they had to stop and lighten up in green (Fig. 1). Each robot had to be programmed differently, in order to make it move from its starting position (indicated by colored spots on the ground map, cf. Fig. 1) into the compartments of the battery. Several hints for solving the tasks were hidden in the room, each one tagged with a color marker (yellow, blue or red, corresponding to the robot the hint was referring to). The hints were placed based on the value of the information they contained (i.e., simple hints were easy to discover, while more valuable hints, e.g., the full solution of a program code, were well hidden). The players were instructed about the game rules beforehand (i.e., no use of violence, no manipulation of the battery, no placing of the robots by hand), but they were not informed about the operating principle of the battery. The winning state had to be discovered by themselves, supported by various notes hidden in the room. Moreover, each robot already comprised an initial behavior at the moment it was found: for instance, putting the blue robot on a black line, would make it follow it. However, if the robot was put on the line leading to the right compartment of the battery, it would eventually enter the compartment, shortly lighten up in green, but then turn and leave. In order to win the game, the players had to add modifications to the program code of each robot. A countdown timer with visual effects and suspenseful music was used to convey a crisis situation. Each game session was supervised by an experimenter who took the role of the game master, providing further guidance to players if it was perceived as necessary (see section *Experimental protocol*). If the group succeeded to place the three robots inside the compartments of the battery, a final video was launched, indicating that the task was accomplished successfully.

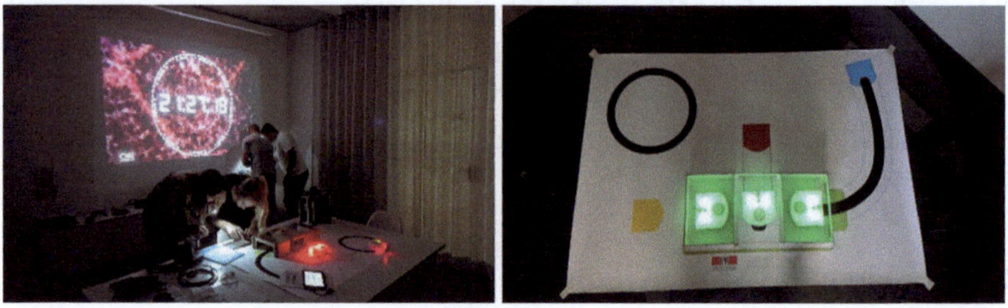

Fig. 1. Game situation (left) and ground map with final winning state (right).

3 Methods

3.1 Participants

The game was tested with 61 participants (31 males, 30 females) of various age groups and different experience in robotics (Table 1). As we aimed at exploring the perception of the game by people of various backgrounds (i.e., by a heterogeneous population), we did not restrict the participation on age or proficiency. However, participants were only included in the study if they had none or limited experience with the Thymio robot and the VPL programming language. For the study, they signed up in groups, i.e., the members of each group knew each other before the experimental session.

Table 1. Demographics and robotics experience of study participants.

Age				Robotics experience			
	Males	Females	Total		Males	Females	Total
Under 15	2	1	3	Novice	6	19	25
15–24	5	14	19	Intermediate	9	8	17
25–44	18	13	31	Advanced	8	3	11
45–64	5	2	7	Expert	8	0	8
Over 65	1	0	1				
Total	31	30	61	Total	31	30	61

3.2 Experimental Protocol

At the onset of an experimental session, the participants were briefly interviewed about their previous experience and affinity to robotics and technology. The information gathered from this evaluation was used by the game master to determine the right balance of guidance during the game. Subsequently, the participants were brought into the playroom and the game started (see section *The Thymio Escape Game*). In contrast to most commercial escape games, we aimed at a high success rate, assuming that this would elicit a more positive learning and game experience. Therefore, the game master was given the possibility to provide extra guidance to the group, if it was perceived as necessary. However, in order to not interrupt the game flow, assistance was provided in a subtle way (i.e., by short comments) and clues were only given with reference to the hints that could be found in the room. The whole experimental session was followed by a second experimenter, who took the role of an observer to gather qualitative data by direct observation. The observer stayed in the background (i.e., in a separate area of the room, that was not accessible to the group) and did not interact with the players. The observations were recorded using written reports for later analysis. After the game, the participants were asked to complete a questionnaire (see section *Data collection*) and were debriefed for ten minutes. The debriefing, a well-known concept to mentally process experiential learning activities [21, 22], was moderated by the game master. The aim was to facilitate a discussion about the performed actions and their consequences, in order to emphasize the learning objectives of the game. Moreover, the participants had the possibility to ask questions about the Thymio robots, the VPL programming language and the escape game itself.

3.3 Data Collection

Quantitative Data. To conduct our survey, we designed a questionnaire which was based on the one developed by Hou and Chou [16]. In their study, they used an evaluation questionnaire to determine the acceptance of their escape game by the participants. The user acceptance was grounded on two dimensions: perceived usefulness and perceived ease of use. Questions related to these two dimensions have been shown to be a reliable measure for the user acceptance of information technology [23]. In their questionnaire, Hou and Chou also utilized the model proposed by Kiilli [24] to determine the flow state of their participants. For our study we adapted their questionnaire to determine the usefulness, ease of use and flow state perceived by the participants during our escape game. We assumed that participants, and particularly novices, would distinguish between two possible uses of the game: initial exploration and comprehension. Therefore, in the dimension of the perceived usefulness we included questions related to both subdimensions. The final questionnaire (Table 2) comprised eighteen statements related to the following dimensions: perceived usefulness for initial exploration (items EXP1-4), perceived usefulness for comprehension (CPR1-4), perceived ease of use (USE1-3) and perceived flow experience (PFL1-6). For the FLOW statement ("*I experienced a clear flow experience during playing*"), participants had to directly state whether they experienced flow during the game. In order to prevent ambiguities, a description of flow taken from the literature [24] was presented in the questionnaire. The participants had to respond to each statement using a 4-point Likert scale (*strongly disagree/disagree/agree/strongly agree*).

Qualitative Data. Qualitative data was collected by means of three open-ended questions in the questionnaire (Table 2). Moreover, it was collected by means of direct observation made by a second experimenter who followed the whole experimental session (i.e., the introduction, the game and the debriefing). The observations were recorded using written reports and they were subsequently discussed among the experimenters in order to determine the significance of each observation.

Table 2. Items of the questionnaire distributed at the end of the game.

Dimension	Item	Statement
Perceived usefulness for initial exploration	EXP1	The game raised my interest in robotics
	EXP2	The game raised my interest in Thymio and VPL
	EXP3	I think the game allows me to better explore Thymio and VPL compared to conventional methods (books, lectures, etc.)
	EXP4	The game helped me explore Thymio and the programming interface VPL
Perceived usefulness for comprehension	CPR1	The game helped me understand the basic concepts of robotics
	CPR2	I think the game allows me to better understand Thymio and VPL compared to conventional methods (books, lectures, etc.)
	CPR3	The game helped me understand how the sensors and actuators of Thymio work
	CPR4	The game helped me understand the basics of the programming language VPL
Perceived ease of use	USE1	The goal of the game is easy to understand
	USE2	The rules of the game are easy to understand
	USE3	I understand the educational goals of the game
Perceived flow experience	PFL1	I felt in total control of my playing actions
	PFL2	The way time passed seemed to be different from normal
	PFL3	I really enjoyed the playing experience
	PFL4	I found the experience extremely rewarding
	PFL5	The challenge that the game provided, and my skills were at an equally high level
	PFL6	I had total concentration while playing the game
Flow statement	FLOW	I experienced a clear flow during playing
Open-ended questions	Q1	About the game, I liked…
	Q2	About the game, I did not like…
	Q3	General suggestions

4 Results

4.1 Quantitative Results

The analysis of the answers given in the questionnaires showed that overall, most participants were very positive about the escape game experience (Fig. 2). On average, we found high approval for the EXP (84% agreed or strongly agreed), CPR (82%), USE (92%) and PFL (89%) dimensions. Moreover, 89% of the participants agreed or strongly agreed on the statement FLOW. However, we also found that particularly statements related to robotics education (i.e., EXP1 and CPR1) found less approval: 21% and 18% of the participants respectively, disagreed on those statements. Following a deeper analysis, we discovered that these answers were mostly given from people with robotics experience - a population for whom those statements are most likely not suited. From the open-ended answers of the questionnaire, we found that for similar reasons, a few participants did not provide answers to some of the statements in the EXP and CPR dimensions. Similarly, participants who have already heard about Thymio and VPL beforehand or had limited experience with the robot, chose to not answer to the corresponding statements (i.e., EXP2-4).

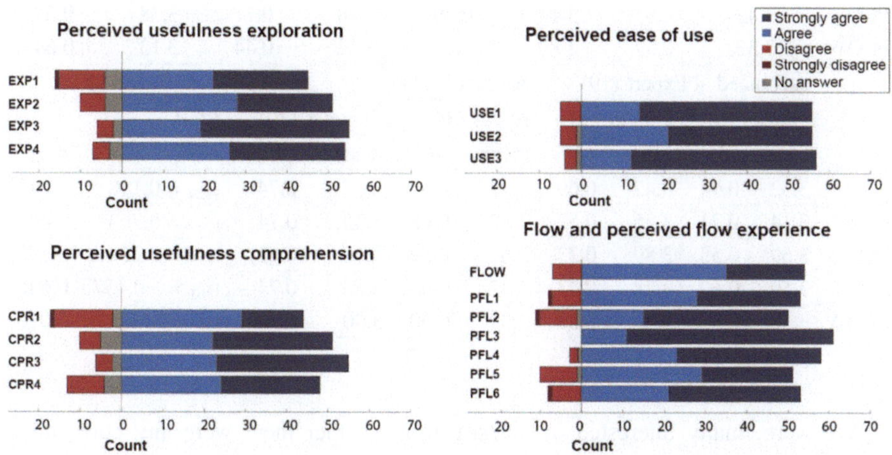

Fig. 2. Summarized results of the questionnaires by each dimension.

A comparison between the groups with different robotics experience (i.e., novices, intermediates, advanced and experts) was performed by analyzing the mean Likert scale values for each dimension. The analysis was done for males and females of each group (Table 3). In general, we observed similar trends for all groups. The only noticeable differences were found for male experts, who, as mentioned before, agreed less on CPR statements than the rest - and more interestingly, also for female novices who agreed less on CPR and FLOW statements. The results illustrated that although the

FLOW item was less approved by females in all groups, it was particularly different for female novices: 32% (6 out of 19) of female novices disagreed on the FLOW item, while only 2% (1 out of 42) of the remaining participants disagreed. Due to the unbalanced group constellations, statistical tests were only performed for gender differences across the different dimensions. A two-sample t-test showed that the only significant difference between males and females was indeed found for the FLOW item ($p = 0.015$). This is in particular interesting, since no significant difference ($p > 0.44$) was found for any of the other four dimensions of the questionnaire, including PFL, which includes statements linked to the flow state.

Table 3. Mean Likert scale values and standard deviation for each dimension (*strongly disagree = 1, disagree = 2, agree = 3, strongly agree = 4*)

Dimension	Novices (25)				Intermediates (17)				Advanced + Expert (19)				All levels (61)					
	Males (6)		*Females (19)*		*Males (9)*		*Females (8)*		*Males (16)*		*Females (3)*		*Males (31)*		*Females (30)*		*t-test*	
	mean	sd	mean	sd	mean	sd	mean	sd	mean	sd	mean	sd	mean	sd	mean	sd	*t*	*p*
EXP	3.38	0.77	3.27	0.79	3.45	0.71	3.52	0.57	3.32	0.54	3.42	0.69	3.37	0.64	3.35	0.74	$t_{229} = 0.198$	0.843
CPR	3.58	0.50	3.15	0.81	3.45	0.67	3.38	0.62	3.14	0.71	3.55	0.52	3.32	0.68	3.25	0.74	$t_{229} = 0.763$	0.446
USE	3.56	0.61	3.54	0.66	3.54	0.64	3.61	0.66	3.69	0.55	3.89	0.33	3.62	0.59	3.60	0.63	$t_{179} = 0.264$	0.792
PFL	3.43	0.65	3.43	0.79	3.28	0.71	3.38	0.68	3.50	0.62	3.78	0.43	3.42	0.66	3.46	0.74	$t_{361} = -0.457$	0.648
FLOW	3.33	0.52	2.89	0.73	3.22	0.44	3.13	0.64	3.50	0.52	3.33	0.58	3.39	0.50	3.00	0.70	$t_{59} = 2.512$	0.015

We were finally interested in investigating whether there were any correlations between the different dimensions. By calculating Pearson's correlation coefficient between the eighteen items of the questionnaire, we found that the strongest correlations appeared for the items within the dimensions of perceived usefulness for exploration and comprehension. Interestingly, we found strong correlations across the two sub-dimensions (i.e., EXP3 with CPR2 and EXP4 with CPR4, $r = 0.75$ for both). Moreover, more than half (9 out of 16) of the correlations between the statements of the EXP and CPR dimensions were at least moderate ($r > 0.42$). These unexpected results were opposed to our initial assumption that participants would consider initial exploration and comprehension as two separate aspects of the escape game's usefulness. Indeed, the results illustrated that if participants accepted the escape game as educationally valuable, they believed it was useful for both initial exploration and actual comprehension.

4.2 Qualitative Results

Qualitative data was collected by means of three open-ended questions in the questionnaire and by direct observation of a second experimenter present during the game sessions (see section *Data collection*). The open-ended questions were completed by almost half (28 out of 61) of the participants. For the first item Q1, we found that participants mostly liked "*the atmosphere of the game*". This feedback, together with the vivid group dynamics we observed, reinforced the results from our quantitative data, indicating that most participants experienced flow. Indeed, we noticed that group members actively shared their discoveries, lively exchanged ideas and verbalized satisfaction: "*Yes we found it!*" (a player after finding a robot in a locked suitcase), "*Ah, now I understand! Look...*" (a player explaining a program code to another), "*Good job! Now the next one!*" (a player after one robot successfully entered the battery). However, we also observed that the dynamics were not the same for all groups. Especially groups with members of varied level of technical proficiency seemed to be less communicative. In these cases, members with lower proficiency appeared to be more passive and rather focused on non-technical tasks, e.g., searching for hints, while the "experts of the groups" dealt with the robots. For the items Q2 and Q3 of the questionnaire we received that many participants desired a more self-contained game, with more hints provided by the game itself rather than having active guidance by the game master. Moreover, some participants felt that the game was too short: "*A little short if the objective is to understand the basics of robotics*", "*To fully understand the code, more time is needed. But the objective of discovery is very good and makes you want to go further*". Although a more passive game master was desired during the actual quest, his role as an active facilitator during the debriefing was highly appreciated. For many participants the debriefing was an eye-opener, allowing them to reflect on their actions and better comprehend the functioning of the robots and the program codes. One participant summarized this observation by the following verbalization: "*After the debriefing, I completely changed my mind about the usefulness of this activity*".

5 Discussion

By introducing the concept of escape games to educational robotics, we sought to determine the experience perceived by the players and the educational potential of such a framework. Based on the two categories of research questions presented in the introduction of this study, we here discuss the results obtained from our experiments.

5.1 Players' Perception of the Game

From the viewpoint of entertainment, the Thymio Escape Game was a great success, since all participants (61 out of 61) stated to have "*really enjoyed the game experience*". This was further supported by qualitative data obtained from direct observations, which illustrated high engagement and enjoyment of the participants during the activity. Moreover, a great majority of the participants agreed on the game's ease of use and its usability as a teaching tool. Unexpectedly, we found strong correlations between EXP and CPR items, refuting our initial assumption that participants would distinguish the game's usability. In contrast, the results showed that if participants found that the game was a valuable teaching tool, they believed it could be useful for both initial exploration and comprehension. However, the results also demonstrated that a positive game and learning experience was not sufficient to ensure a flow experience for all participants. Indeed, 11% (7 out of 61) of the participants disagreed on having "*experienced a clear flow during playing*". We found a significant difference for the FLOW statement across genders: while all males (100%) agreed or strongly agreed on having experienced flow during the game, this was true for only 73% of the females. This observation is in line with the results of Hou and Chou [16], who reported that males experienced more flow than females. However, when considering the statements of the PFL dimension, we observed that the perceived flow experience was similar for both genders. We presume that this result could be related to the presentation of the FLOW statement, which in our case may not have been sufficient to fully capture the meaning of flow. Moreover, we found that the major contribution to the disparity in the FLOW statement was caused by female novices. Indeed, 6 out of the 7 participants who disagreed on the FLOW statement, belonged to this group. Looking at the gender group sizes for each cluster based on robotics experience, we found that females were predominant in the novices' group. Therefore, another way to interpret these results could be that disparities in the FLOW statement were actually not based on gender differences, but on the proficiency level of the participants. However, further inquiries with larger and more balanced sample sizes would be needed to draw meaningful conclusions.

5.2 Educational Potentials of Escape Games

As presented in the previous section most participants perceived the escape game as educationally valuable. Indeed, our observations illustrated that it provides a framework in which people strongly collaborate to achieve a common goal. We observed that participants were self-driven to understand the functioning of the setup and lively exchanged ideas and knowledge among each other. Although the game master was present to facilitate the experience, the participants' motivation to comprehend and solve the given problem situation was highly intrinsic. Indeed, the answers given in the questionnaire indicated the desire for a more self-contained game, imposing a less active role on the game master and more autonomy for the players. These results strongly support our initial idea, that escape games could be a favorable instructional approach in the field of ER to put self-regulated and collaborative learning into practice. In contrast to other constructivist approaches, escape games provide a well-framed environment, which allows the learners to autonomously explore and construct knowledge without having to allocate cognitive resources to activities that are not related to the learning goals, alleviating one main criticism on constructivism found in literature [25]. The well-defined common goal (i.e., reaching the winning state), that can only be achieved following a pre-determined sequence of tasks, not only provides a more structured framework to implement constructivist methodologies – it also imposes a positive social interdependency among the players, a condition known to be beneficial for collaborative learning scenarios [26]. The possibility to give hints, whether indicated by clues in the room or communicated by the game master, additionally provides means to implement and continuously adjust instructional scaffolding. We, therefore, believe that the presented perspectives on escape games may not only apply in the context of ER but can potentially be extended to many other disciplines. However, we suggest that ER may particularly be suited for the integration of escape game activities, due to its inherent constructivist roots [27]. The capability to trigger high intrinsic motivation in both genders and at all proficiency levels, could make it a valuable teaching tool for ER. However, whether escape games are a more efficient way of instruction, still has to be demonstrated. Future work could therefore assess the learning outcomes gained from escape games in comparison to conventional methods. In this context, the group composition, as in any collaborative activity, may have an influence on the game and the learning experience. Previous works have investigated collaboration in commercial escape games [28] and likewise, the effect of the group composition in collaborative learning has been discussed before [29]. Bringing together these evidences could thus further improve the learning experiences induced by educational escape games.

6 Conclusion

In this work we presented a prototype of an escape game, which was developed in the context of educational robotics. Based on quantitative and qualitative results from experiments with 61 subjects, we found that participants highly appreciated the game experience and agreed on the game's usability for teaching. We, therefore, conclude that escape games have auspicious potentials as a teaching tool in educational robotics. The underlying structure of escape games provides a favorable framework to put self-regulated and collaborative learning into practice, hence optimally exploiting the inherent constructivist character of educational robotics. By incorporating a captivating and joyful environment, escape games can elicit high intrinsic motivation for learning, and hence potentially provide means for more purposive and efficient instruction.

References

1. Deci, E., Vallerand, R., Pelletier, L., Ryan, R.: Motivation and education: the self-determination perspective. Educ. Psychol. **26**, 325–346 (1991)
2. Vallerand, R.J., Pelletier, L.G., Blais, M.R., et al.: The academic motivation scale: a measure of intrinsic, extrinsic, and amotivation in education. Educ. Psychol. Meas. **52**, 1003–1017 (1992)
3. Viau, R.: La motivation en contexte scolaire. De Boeck Supérieur (2009)
4. Prensky, M.: Computer games and learning: digital-based games. In: Handbook of Computer. Game Studies, pp. 97–124 (2005)
5. Fletcher, J.D., Sigmund, T., Wind, A.P.: Game-based learning. In: Handbook of Research on Educational Communications and Technology, 4th edn., pp. 485–503 (2014)
6. Whitton, N.: Digital Games and Learning: Research and Theory (2014)

7. Steinkuehler, C., Squire, K.: Videogames and learning. In: The Cambridge Handbook of the Learning Sciences, pp. 377–394 (2014)
8. Clarke, S., Peel, D.J., Arnab, S., et al.: escapED: a framework for creating educational escape rooms and interactive games for higher/further education. Int. J. Serious Games **4**, 73–86 (2017)
9. Nicholson, S.: Peeking behind the locked door: a survey of escape room facilities. White Paper, pp. 1–35 (2015). http://scottnicholson.com/pubs/erfacwhite.pdf
10. Vygotsky, L.S.: Mind in Society: The Development of Higher Psychological Processes (1978)
11. King, A.: From sage on the stage to guide on the side. Coll. Teach. **41**, 30–35 (1993)
12. Papert, S.: Situating constructionism. Constructionism **36**, 1–11 (1991)
13. Piaget, J.: The construction of reality in the child. J. Consult. Psychol. **19**, 77 (1955)
14. Vörös, A.I.V., Sárközi, Z.: Physics escape room as an educational tool. In: AIP Conference Proceedings, vol. 1916 (2017)
15. Kinio, A., Dufresne, L., Brandys, T., Jetty, P.: Break out of the classroom: the use of escape rooms as an alternative learning strategy for surgical education. J. Vasc. Surg. **66**, e76 (2017)
16. Hou, H.-T., Chou, Y.-S.: Exploring the technology acceptance and flow state of a chamber escape game - escape the lab© for learning electromagnet concept. In: Work Poster Proceedings of 20th International Conference on Computers in Education, ICCE 2012, pp. 38–41 (2012)
17. Shernoff, D.J., Csikszentmihalyi, M., Shneider, B., Shernoff, E.S.: Student engagement in high school classrooms from the perspective of flow theory. Sch. Psychol. Q. **18**, 158–176 (2003)
18. Chevalier, M., Riedo, F., Mondada, F.: Pedagogical uses of Thymio II. IEEE Robot. Autom. Mag. **23**, 16–23 (2016)
19. Boller, S., Kapp, K.M.: Play to Learn: Everything You Need to Know About Designing Effective Learning Games. ATD Press, Alexandria (2017)
20. Shin, J., Siegwart, R., Magnenat, S.: Visual programming language for Thymio II robot. In: Conference on Interaction Design and Children (2014). https://doi.org/10.3929/ethz-a-010144554
21. Dennehy, R.F., Sims, R.R., Collins, H.E.: Debriefing experiential learning exercises: a theoretical and practical guide for success. J. Manage. Educ. **22**, 9–25 (1998)
22. Nicholson, S.: Completing the experience: debriefing in experiential educational games. In: 3rd International Conference on Society and Information Technologies, vol. 11, pp. 117–121 (2012)
23. Davis, F.D., Davis, D.: Perceived usefulness, perceived ease of use, and user acceptance of information technology. MIS Q. **13**, 319–340 (1989)
24. Kiili, K.: Evaluations of an experiential gaming model. Hum. Technol. Interdisc. Hum. ICT Environ. **2**, 187–201 (2006)
25. Kirschner, P.A., Sweller, J., Clark, R.E.: Why minimal guidance during instruction does not work. Educ. Psychol. **41**, 87–98 (2006)
26. Johnson, D.W., Johnson, R.T.: Assessing Students in Groups, 1st edn. Corwin (2004)
27. Alimisis, D.: Educational robotics: Open questions and new challenges. Themes Sci. Technol. Educ. **6**, 63–71 (2013)
28. Pan, R., Lo, H., Neustaedter, C.: Collaboration, awareness, and communication in real-life escape rooms. In: Proceedings of 2017 Conference on Designing Interactive Systems - DIS 2017, pp. 1353–1364 (2017)
29. Slavin, R.E.: Research on cooperative learning and achievement: what we know, what we need to know. Contemp. Educ. Psychol. **21**, 70–79 (1995)

Teaching Robot Programming for Industry 4.0

Elisa Tosello[1(✉)], Nicola Castaman[1,2], Stefano Michieletto[1], and Emanuele Menegatti[1]

[1] Department of Information Engineering, University of Padova, Padova, Italy
{toselloe,castaman,michieletto,emg}@dei.unipd.it
[2] IT+Robotics srl, Vicenza, Italy

Abstract. This paper presents a master course project on intelligent robotics offered by the University of Padova (Italy). The goal is that of training Master students for Industry 4.0 by offering a multidisciplinary laboratory experience in which two robots and a human must collaborate to fulfill an assembly task: one manipulator robot has to recognize some pieces on a table, manipulate them, and place them on the top of a mobile robot. The mobile robot has to carry these pieces within an arena until reaching an assembly station where the human operator will complete the assembly task. Through our constructivist approach, students learn how to face a rapidly evolving discipline requiring the integration and cooperation of multiple subsystems to create complex behaviors. They learn how to program autonomous robots by using the Robot Operating System. They learn how to exhaustively document their activity explaining design choices, the benefits and the limits of their approaches, as well as proposing new solutions to overcome these limitations. They learn how to properly implement and comment code as well as create step-by-step instructions to enable future users to reproduce their systems. The project is organized as a challenge to motivate students to propose innovative ideas: students are subdivided into teams, every team proposes its own solution, and different scores are assigned according to the proposed difficulty level and the required working time.

Keywords: Educational robotics · Industry 4.0 · Robot Operating System · Robot-robot cooperation · Human-robot cooperation

1 Introduction

As many economists and policy makers are reporting, the fourth industrial revolution, namely Industry 4.0 [1], is underway. It aims to develop "*Smart Factories*": digitized manufacturing companies equipped with Cyber-Physical Systems (CPSs)[1]. CPSs are complex systems controlled by computer-based algorithms and equipped with physical and software components able to interact with each

[1] See https://www.nsf.gov/funding/pgm_summ.jsp?pims_id=503286.

M. Moro et al. (Eds.): Edurobotics 2018, AISC 946, pp. 107–119, 2020.
https://doi.org/10.1007/978-3-030-18141-3_9

other, perceive their surroundings, and reason on assigned tasks. These systems operate within the production chain but do not slavishly repeat the same pre-programmed routines. They adapt their operations by communicating and collaborating with each other and with human operators. Keeping in mind this new way of working, the new generation of engineers should not be expert on one single domain of knowledge. They should be able to integrate multidisciplinary and cross-domain knowledge. They should cope with new paradigms and concepts (e.g., modeling, simulation, interoperability and self-organization) and emergent technologies (e.g., IoT and Artificial Intelligence) [2].

The laboratory activity reported in this paper exactly focuses on this multi-disciplinary challenge: it asks students to program two different robots in order to fulfill an assembly task in collaboration with a human operator. This means that students do not only have to face classical robotics problems such as 3D perception, mobile navigation or manipulation planning. They have to get in perspective of one single complex task that has to be subdivided into simpler sub-tasks, each of them has to be assigned to one single robot depending on its capabilities. This complex task is then achieved thanks to an efficient robot-robot and human-robot collaboration. The laboratory activity is part of "*Autonomous Robotics*" (AR): a second year course of the Master of Science (MSc) in "*Computer Science*" at the School of Engineering of the University of Padova (Italy). Students of "*Computer Science*" can also take a second robotics course: "*Robotics, Vision and Control*". This course is offered by the same School but within the degree of Automation Engineering. It takes place in the semester next to that of AR and, as AR, it is optional. The two courses are scheduled in order to offer different skills. AR aims to increase students' programming skills and their theoretical background on intelligent, autonomous systems (e.g., how to make a robot solve task and motion planning tasks, how to make it reason). The second course focuses on robot control.

We want to highlight that the proposed activities come from a multi-year experience: the lab assignment proposed in this paper replaced the ones proposed in previous years. Before the Indutry 4.0 challenge, students had to solve six homeworks, homogeneously assigned throughout the duration of the course [3,4]. Every lab asked students to face one single robotics problem by means of one robot: among others, we asked to solve a navigation problem by means of a Lego Mindstorms NXT robot[2], a manipulation problem by means of a Universal Robots UR10[3], and a teleoperation problem by means of a humanoid Robovie-X robot[4]. The limit that pushed us on proposing one single complex project is the lacking of a global view on how robotics systems should cooperate to fulfill complex tasks.

The rest of the paper is organized as follows. Section 2 shows related works on this field. Section 3 gives an overview of the "*Autonomous Robotics*" course describing the content of both the lectures and the laboratory assignment,

[2] See https://www.lego.com/it-it/mindstorms.
[3] See https://www.universal-robots.com/products/ur10-robot/.
[4] See https://www.vstone.co.jp/english/products/robovie_x/.

together with the prerequisites needed to take the course. Sections 4 and 5 detail the laboratory assignment. First, the intended learning outcomes are depicted. Then, both the hardware and software setups are given, together with the programming tasks assigned to students. Section 6 analyzes best solutions proposed by students. Section 7 contains results in terms of students satisfaction. Finally, Sect. 8 contains conclusions and future work.

2 State of the Art

There exists both long and short-term academic learning programs that provide a robotics multidisciplinary specialization. However, we do not aware of long-term ones that focus on Industry 4.0. An example is Duckietown [5]: even if this project still aims to educate engineers able to respond to current market needs, it focuses on autonomous cars. Focusing on short-term programs, instead, [2] reminds the Erasmus Intensive program on Robotic Systems [6] and the Summer School on Industrial Agents in Automation [7]. The former is a 60 h course which provides global insight knowledge on robotic systems and hands-on experience in this field. It introduces concepts and technologies for different domain applications, such as industrial robotics, autonomous mobile robotics and robotics applied to medicine. It integrates knowledge from six European High-level Education Institutions from Portugal, Spain, United Kingdom, Germany and Turkey. The latter is a 28 h course focused on Multi-Agent Systems applied to industrial environments. Another initiative is the Robot Operating System (ROS) Summer School[5]. ROS [8] is a robotics middleware that provides software libraries for solving multiple robotics tasks such as mobile navigation, robot manipulation, and perception. The summer school lasts two weeks and provides both an introduction to the ROS most important packages and a deep overview of mobile robot routines, i.e., perception, localization, and navigation. A competition is scheduled at the end of course in which participants are subdivided in teams and every team has to design a mobile robotics application such as indoor/outdoor exploration. They all use the same hardware but have to implement the correct ROS software routines. Other competitions exist to bring students closer to robotics. Examples are the ROCKIn@Home [9] and @Work [10] challenges, the RoboCup@Home and @Work challenges[6], the Mohamed Bin Zayed International Robotics Challenge (MBZIRC)[7] and the European Robotics Challenges (Euroc)[8].

Our Master course mixes benefits of competitions with the requirements of multidisciplinary teaching in the context of Industry 4.0. The proposed scenario takes inspiration from the Amazon Picking Challenge [11] and, like in all challenges, a small competition is organized for those students who want to present

[5] See https://www.fh-achen.de/fachbereiche/maschinenbau-und-mechatronik/international/ros/.

[6] See http://www.robocup-2015.org/index.htmt.

[7] see https://www.mbzirc.com/.

[8] See http://www.euroc-project.eu/.

their solutions exactly at the end of the course. Its attendance is not mandatory but winners gain two more points on their final mark. From our RoboCup [12], MBZIRC [13] and Euroc [14] experiences, we realized that proposing a competition motivates students to present innovative solutions and to take the exam by the end of the course. Moreover, as current trends impose, we ask students to program by using ROS. Indeed, ROS is open source, it is fine grained and consists of numerous reusable modules. Thus, students learn how to organize software into modules, reuse data structure and classes, and exploit class inheritance [15]. ROS also provides tools and libraries for obtaining, building, writing, and running code across multiple computers with a powerful communication protocol. Thus, students can easily implement a network of robots (and computers) able to exchange data and interact with each other.

3 The Course

"Autonomous Robotics" (AR) is a second year course of the Master of Science (MSc) in "Computer Science" at the School of Engineering of the University of Padova (Italy). It lasts 12 weeks and offers students methodological bases for programming autonomous robotic systems by combining class lectures and a laboratory assignment. While lectures give students a theoretical background on robotic fundamentals, the lab lets students apply these fundamentals for the resolution of a robotic task. Students are examined both on lectures (30% of the final grade) and on the programming assignment (the remaining 70%). As in competitions, the assignment is subdivided into sub-tasks and a scoring schema is made known to students, assigning a specific score to each sub-task.

3.1 Prerequisites and Skills

No prerequisite is mandatory for course attendance. In class, students acquire both the theoretical and programming skills necessary to face both the final written test and the lab assignment. However, having basic C++ programming capabilities facilitates the implementation of the lab. Indeed, ROS accepts both C++ and Python commands but we suggest the use of C++ because a course on C++ programming is offered within our same course of study.

3.2 Lectures

Three classes of two hours per week are taught. They give an introduction of both mobile and manipulator robots with a deep overview on the relationship between perception and action in robotic systems, existing algorithms for solving motion planning problems, and learning methods for the autonomous execution of assigned tasks. During class lessons, five tutorials of two hours each introduce students to ROS programming. Tutorials give an overview of ROS architecture and how to implement publishers, subscribers, actions and services.

Fig. 1. Lab project: the toy assembly task.

They teach how to use the ROS navigation stack[9], MoveIt![10], and how to implement advanced perception routines by means of the AprilTags Visual Fiducial System [16] and the Point Cloud Library (PCL) [17].

3.3 Lab Project

The Industry 4.0 laboratory experience has been designed as follows: multiple different objects are on a table in front of a UR10 manipulator robot, a Microsoft Kinect v2 is mounted on this table. The robot has to detect these objects, pick them up, and place them on the top of a Turtlebot 2 robot[11] previously docked next to the UR10. Turtlebot 2 has to navigate inside an arena (see Fig. 2) and reach a second docking area where one student has to take the objects and assemble a simple construction (see Fig. 1). Students have to cope with the following constraints:

(a) A timeout is imposed (30 min).
(b) Pieces are subdivided into three categories, each with peculiar shape, color, and weight (red cubes, green triangles, and yellow prisms). Every category is associated with scores proportional to the grasping difficulty.
(c) The area on the top of the mobile robot is limited: students have to implement an optimal policy to transport all pieces in the minimum number of travels and to maximize their scores.
(d) The mobile robot has to reach fixed positions in the map (docking areas) in order to have the objects loaded by the manipulator or unloaded by the human operator.
(e) The arena is populated by both fixed and movable obstacles.
(f) The mobile robot always starts from a prefixed pose.

---

[9] See http://wiki.ros.org/navigation.
[10] See http://moveit.ros.org.
[11] See https://www.turtlebot.com/.

Students have to face the following additional challenges:

(a) Turtlebot 2 has to go through a narrow passage 5 cm larger than the robot.
(b) Objects should be recognized and their poses should be precisely estimated: students can exploit both the fiducial markers attached on the top of the objects and the point clouds acquired by the Kinect.
(c) To accomplish the task, robots should coordinate with each other: students should implement a Finite State Machine to control the task execution.

4 Lab Project: Educational Details

4.1 Teaching Methodology

A constructivist approach is the basis of our laboratory assignment. It combines project-based, team-based, and peer-based learning as follows.

(a) *Project-based learning*: a complex, single project is assigned to students;
(b) *Team-based learning*: students organize themselves in teams of two/three people (classes of about 20 students per year) and every team has to propose its own solution. Performed tests demonstrate that having larger groups induce confusion and unbalanced workload division within the group itself;
(c) *Peer-based learning*: students have to present their own solutions within three months after the end of the course. At the exam, they must provide both a demonstration proving the correct functioning of their system and an oral presentation exposing which algorithms they adopted to achieve their results. During the presentation, students should highlight both benefits and limits of their approaches and they should propose ideas to overcome these limitations. For those students who want to present the project exactly at the end of the course, a challenge is organized. The winning team gains two more points on the final grade. In our opinion, facing a challenge motivates students to present innovative solutions and take the exam by the end of the course;
(d) *Constructivist learning*: in order to solve the assignment, students can access the lab and use available robots every day. One teaching assistant is in the lab to supervise and help students. No more teachers are required because students should not be slavishly guided until the fulfillment of assigned tasks. They should act by themselves: during the class, they become aware of the assignment, they discuss what algorithms to use to solve the problem, and schedule their lab activity. In the lab, they try to implement their choices, they discuss what improvements are feasible and they work together to make these adjustments.

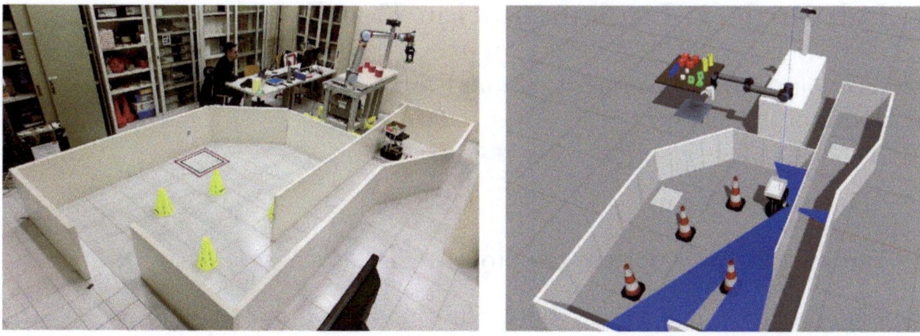

(a) The real-world arena. (b) The simulated arena.

Fig. 2. Lab project: the arena.

4.2 Intended Learning Outcomes

As for [5], we divided the Intended Learning Outcomes (ILOs) into four categories: *operational tools*, *development methods*, *autonomy* (perception, control, coordination) and *documentation*. By the end of the course, students should be able to:

(a) Use *operational tools*, included Ubuntu and ROS.
(b) Acquire knowledge on *development methods*, such as:
  - (DEV-ROS) develop ROS software modules and integrate them in their systems;
  - (DEV-Tools) utilize standard tools for software development (e.g., source code repositories; branching and merging);
  - (DEV-Group) familiarize with the dynamics of open-source development, including the challenge of integrating independently-developed functionalities.

(c) Implement features guaranteeing robots *autonomy* by:
  - (AU-Perception) guaranteeing the autonomous detection and recognition of manipulation objects;
  - (AU-Control and Motion Planning) guaranteeing the autonomous manipulation and navigation. Collisions should be avoided and the mobile robot should be able to autonomously face the narrow passage;
  - (AU-Coordination) guaranteeing the autonomous coordination and cooperation of available robots.

(d) Write an exhaustive *Documentation* by:
  - (DOC-Explain) preparing a presentation with design choices, system benefits, limitations and ideas on how to overcome these limitations;
  - (DOC-Document) commenting their code and creating step-by-step instructions to enable future users to reproduce their systems.

5 Lab Project: Implementation Details

This Section describes the hardware and software setups needed to reproduce our laboratory assignment.

5.1 Hardware Setup

The hardware available for the lab project (Fig. 2a) consists of:

(a) a lightweight collaborative robot, the Universal Robots UR10;
(b) a three finger gripper, the Robotiq 3-Finger Gripper[12];
(c) a mobile platform, the Turtlebot 2 provided by Robotnik;
(d) a RGB-D camera, the Microsoft Kinect v2;
(e) a Laser Range Finder (LRF), the Hokuyo URG-04LX-UG01[13];
(f) a webcam, the Philips SPZ5000;
(g) an embedded platform, the NVIDIA Jetson TX1[14];
(h) a small laptop, the Lenovo Thinkpad X201;
(i) a wireless router, the TP-Link Archer C7[15];
(j) a network switch, the TP-Link TL-SG108;
(k) several network cable to connect students' PCs to the project network.

The UR10 is the biggest model of the Universal Robots family of manipulator. It is a collaborative industrial robotic arm and it can work alongside humans with no need for fences. The reach radius of 1.3 m and the maximum payload of 10 Kg make this arm suitable for the lab project activities. The robot can easily reach both the objects on the table and the destination box on the mobile platform. The UR10 has been equipped with the Robotiq 3-Finger Gripper. This adaptive gripper provides high flexibility and versatility in the object picking task, due to the 4 grip types available: (i) Pinch Mode, (ii) Wide Mode, (ii) Scissor Mode, and (iv) Basic Mode. The Microsoft Kinect v2 can see objects on the table from an external and fixed point of view. A 2D image, in FullHD resolution, and a depth image, at 512×424 pixels, are streamed from the camera sensor at 30 Hz, along with the intrinsic and extrinsic parameters for the integrated sensors. Students can take advantage of the necessary information depending on the approach they are using to identify the 6D pose of the objects in the scene. TurtleBot 2 is an open-source robot built upon the Kobuki mobile base, in a modular fashion. A set of plates can be mounted on the top of its base letting us obtain the best configuration for our purpose. For the lab project, we disassembled the Microsoft Kinect v1 from the Turtlebot hardware and replaced it with an Hokuyo URG-04LX-UG01 and a Philips SPZ5000. The LRF is more reliable in mapping the environment, especially when sunlight enters from the windows. Moreover, the

[12] See https://robotiq.com/products/3-finger-adaptive-robot-gripper.
[13] See https://www.hokuyo-aut.jp/search/single.php?serial=166.
[14] See https://www.nvidia.com/en-us/autonomous-machines/embedded-systems-dev-kits-modules/.
[15] See https://www.tp-link.com/it/products/details/cat-9_Archer-C7.html.

webcam produces a higher resolution image with respect to the first version of the Microsoft camera. We mounted a box on the top of the platform for containing and carrying around the objects collected by the robotic arm. All devices are connected together forming both a cabled and a wireless network by means of a TP-Link TL-SG108 switch and a TP-Link Archer C7 router, respectively. The UR10 and the Robotiq 3-Finger Gripper have been directly plugged into the network. The Microsoft Kinect v2 has been connected through the NVIDIA Jetson TX1 board while the Turtlebot 2 platform and the sensors mounted on it can communicate via the Lenovo Thinkpad X201 notebook.

5.2 Software Setup

The project is based on ROS. We adopted Ubuntu 16.04 and ROS Kinetic: the last LST version at the time of the project formulation.

A simulated version of the real-world setup is provided (see Fig. 2b). It is absolutely identical to its real counterpart and lets students test the efficiency and correctness of their designs. Students can use simulation to understand how their systems will act in real world and how to correct failures. Gazebo is used as simulator [18] as it is the official ROS simulator.

Multiple packages are provided in order to help students face the project:

(a) The Turtlebot 2 ROS packages and the ROS navigation stack are provided to control the mobile platform and move it inside the arena while avoiding collisions.
(b) The ROS-Industrial `universal_robot`[16] and `robotiq`[17] packages are provided to control the UR10 and the gripper, respectively. We substituted the Universal Robot driver with the so called UR Modern Driver[18][19]: an optimized UR driver from Thomas Timm Andersen;
(c) The Kinect sensor has to be used for object identification and pose recognition. We provide the IAI Kinect2[19] library. This package contains a calibration tool for calibrating the IR sensor with respect to the RGB sensor and depth measurements, a library for depth registration with OpenCV support [20], and the bridge between `libfreenect2`[20] and ROS.
(d) Two techniques are proposed for object recognition: the first one is based on fiducial markers and the other one requires the point cloud elaboration. We suggest AprilTag as fiducial marker system. In the preliminary tests we made, it resulted more accurate with respect to other systems, such as ARTag. The AprilTags C++ library lets the detection of AprilTags in a given image and provides the unique ID of the tag as well as its location in the image. There exists many ROS AprilTags wrappers, we made RIVeR-Lab[21] available. Object recognition based on object features (e.g., shapes,

[16] See https://github.com/ros-industrial/universal_robot.
[17] see https://github.com/ros-industrial/robotiq.
[18] See https://github.com/ThomasTimm/ur_modern_driver.
[19] See https://github.com/code-iai/iai_kinect2.
[20] See https://github.com/OpenKinect/libfreenect2.
[21] See https://github.com/RIVeR-Lab/apriltags_ros.

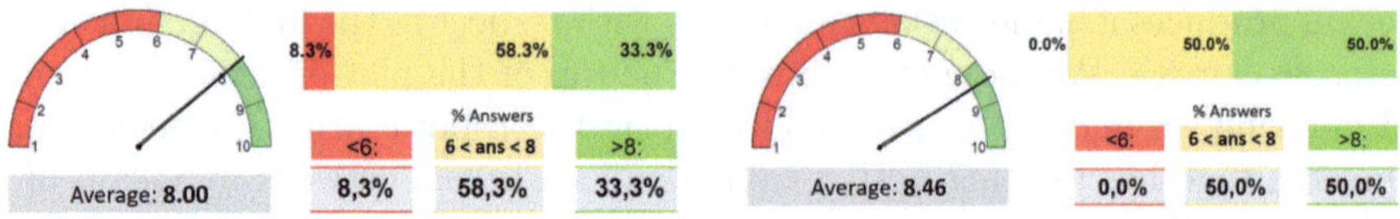

(a) Overall satisfaction.

(b) Indicator of the educational action.

Fig. 3. Results of the survey on students' opinions.

colors), instead, requires the point cloud elaboration. PCL can be used to process 2D/3D images and point clouds and its ROS interface is already provided by the ROS installation;

All the packages and the simulation setup are distributed to students through Git. In this way, it is easy to solve bugs and distribute software updates.

5.3 Programming Assignments

Students have to implement a Finite State Machine able to coordinate robots while they are accomplishing their own sub-tasks. Students have to implement the perception, manipulation, and navigation routines to solve the assigned task at their best, by exploiting the provided ROS packages or by developing their own software. Every programming effort is equally scored.

6 Proposed Solutions Analysis

In A.Y. 2017/2018, teams fulfilled the laboratory assignment in different ways and with different strategies. Below, we report the antipodal solutions proposed by the two best teams.

One team completed the lab choosing the minimum difficult coefficient in favor of a more reliable and faster solution. For object detection, they exploited the AprilTag markers on the top of the objects, MoveIt! was used to plan the manipulation, and the ROS navigation stack was used for the navigation. They accurately turned this stack parameters to avoid collisions. To go through the narrow passage, they combined known information, e.g., tunnel width and geometry, with laser data and implemented a routine letting the robot travel a path equidistant from the passage walls. Using this approach, their software was ready at the end of the course, they faced the challenge and succeeded at their first run in about 20 min.

The other best team developed the most innovative solution collecting the maximum difficulty coefficient multiplier, but they needed 3 additional months to develop their solution. During the trial some penalties were found due to imprecise robot movements. They exploited objects' colors for object recognition. Again, MoveIt! was used for solving the manipulation problem but an ad-hoc routine was implemented for the navigation part. This routine exploited potential fields to face both obstacle avoidance and narrow passage.

These results confirm the lessons learned by the Desert Lion team of our lab during the MBZIRC Challenge [13]: while facing robotics challenges, a good trade-off has to be found between a reliable, fast to implement, solution and an innovative, efficient, but maybe unstable, one.

7 Students Satisfaction

At the end of the A.Y. 2017/2018 course, students completed an anonymous questionnaire on what they liked/disliked of the course and what they suggested to change. 9 out of 16 students completed the questionnaire. Obtained answers confirm that the proposed course activity was appreciated. In Fig. 3a, the overall students satisfaction is reported. Students assigned an average grade of 8.0 to the course. This result highlights students overall satisfaction on how the course was held. In Fig. 3b, the indicator of the educational action is depicted. An average grade of 8.46 was assigned, highlighting that the teacher correctly stimulated and motivated the interest in the robotics discipline, beyond clearly explaining the course content. For completeness of information, 50% of students said that they would have needed more knowledge of C++ programming. Moreover, 37.5% complained that the lab required to much effort.

8 Conclusions and Future Work

In this paper, we presented the laboratory project proposed by the "*Autonomous Robotics*" course of the M.Sc. degree in "*Computer Science*" at the School of Engineering of the University of Padova (Italy). The project asks students to ROS program two different robots (a UR10 manipulator robot and a Turtlebot 2 mobile robot) to cooperate for object picking and transportation. The project is assigned to students in terms of challenge and wants to train students for Industry 4.0. A constructivist approach is adopted based on a project-based, team-based, peer-based learning. The effectiveness of our proposal is proved by students' satisfaction through an anonymous questionnaire. In the future, we plan to continue the students' approach to Industry 4.0 by introducing a more effective human-robot interaction and the Cloud support.

References

1. Kagermann, H., Wahlster, W., Helbig, J.: Recommendations for implementing the strategic initiative industrie 4.0 – securing the future of German manufacturing industry. Final report of the industrie 4.0 working group, acatech – National Academy of Science and Engineering, April 2013
2. Leito, P.: Engineering education efforts to support industry 4.0. In: CLME2017/ VCEM - 8 Congresso Luso-Moambicano de Engenharia/V Congresso de Engenharia de Moambique, September 2017

3. Michieletto, S., Tosello, E., Pagello, E., Menegatti, E.: Teaching humanoid robotics by means of human teleoperation through RGB-D sensors. Robot. Auton. Syst. **75**, 671–678 (2016)
4. Tosello, E., Michieletto, S., Pagello, E.: Training master students to program both virtual and real autonomous robots in a teaching laboratory. In: 2016 IEEE Global Engineering Education Conference, EDUCON 2016, Abu Dhabi, United Arab Emirates, 10–13 April 2016, pp. 621–630 (2016)
5. Paull, L., Tani, J., Ahn, H., Alonso-Mora, J., Carlone, L., Cap, M., Chen, Y.F., Choi, C., Dusek, J., Fang, Y., Hoehener, D., Liu, S., Novitzky, M., Okuyama, I.F., Pazis, J., Rosman, G., Varricchio, V., Wang, H., Yershov, D., Zhao, H., Benjamin, M., Carr, C., Zuber, M., Karaman, S., Frazzoli, E., Vecchio, D.D., Rus, D., How, J., Leonard, J., Censi, A.: Duckietown: an open, inexpensive and flexible platform for autonomy education and research. In: 2017 IEEE International Conference on Robotics and Automation (ICRA), pp. 1497–1504, May 2017
6. Leito, P., Fraile, J.C., Moreno, V., Harrison, R., Altun, H., Colombo, A.W., Turiel, J.P., Curto, B.: Transnational lifelong education course in robotic systems. In: IECON 2015 - 41st Annual Conference of the IEEE Industrial Electronics Society, pp. 004181–004186, November 2015
7. Leito, P., Ribeiro, L., Barata, J., Vogel-Heuser, B.: Summer school on intelligent agents in automation: hands-on educational experience on deploying industrial agents. In: IECON 2016 - 42nd Annual Conference of the IEEE Industrial Electronics Society, pp. 6602–6607, October 2016
8. Quigley, M., Conley, K., Gerkey, B.P., Faust, J., Foote, T., Leibs, J., Wheeler, R., Ng, A.Y.: ROS: an open-source robot operating system. In: ICRA Workshop on Open Source Software (2009)
9. Schneider, S., Hegger, F., Ahmad, A., Awaad, I., Amigoni, F., Berghofer, J., Bischoff, R., Bonarini, A., Dwiputra, R., Fontana, G., Hochgeschwender, N., Iocchi, L., Kraetzschmar, G., Lima, P.U., Matteucci, M., Nardi, D., Schiaffonati, V.: ISR. In: The RoCKIn@Home Challenge (2014)
10. Dwiputra, R., Berghofer, J., Ahmad, A., Awaad, I., Amigoni, F., Bischoff, R., Bonarini, A., Fontana, G., Hegger, F., Hochgeschwender, N., Iocchi, L., Kraetzschmar, G., Lima, P.U., Matteucci, M., Nardi, D., Schiaffonati, V., Schneider, S.: ISR. In: The RoCKIn@Work Challenge (2014)
11. Correll, N., Bekris, K.E., Berenson, D., Brock, O., Causo, A., Hauser, K., Okada, K., Rodriguez, A., Romano, J.M., Wurman, P.R.: Analysis and observations from the first amazon picking challenge. IEEE Trans. Autom. Sci. Eng. **15**(1), 172–188 (2018)
12. Nardi, D., Noda, I., Ribeiro, F., Stone, P., von Stryk, O., Veloso, M.: RoboCup soccer leagues. AI Mag. **35**(3), 77–85 (2014)
13. Castaman, N., Tosello, E., Antonello, M., Bagarello, N., Gandin, S., Carraro, M., Munaro, M., Bortoletto, R., Ghidoni, S., Menegatti, E., Pagello, E.: RUR53: an unmanned ground vehicle for navigation, recognition and manipulation. CoRR abs/1711.08764 (2017). http://arxiv.org/abs/1711.08764
14. Michieletto, S., Pagello, E.: Competitions and industrial tasks as a way to learn basic concepts in robotics. In: 2018 IEEE International Conference on Autonomous Robot Systems and Competitions, ICARSC 2018, Torres Vedras, Portugal, 25–27 April 2018, pp. 173–178 (2018)
15. Michieletto, S., Ghidoni, S., Pagello, E., Moro, M., Menegatti, E.: Why teach robotics using ROS? J. Autom. Mob. Robot. Intell. Syst. **8**(1), 60–68 (2014)

16. Olson, E.: AprilTag: a robust and flexible visual fiducial system. In: Proceedings of the IEEE International Conference on Robotics and Automation (ICRA), pp. 3400–3407. IEEE, May 2011
17. Rusu, R.B., Cousins, S.: 3D is here: point cloud library (PCL). In: IEEE International Conference on Robotics and Automation (ICRA), Shanghai, China, 9–13 May 2011 (2011)
18. Koenig, N., Howard, A.: Design and use paradigms for gazebo, an open-source multi-robot simulator. In: IEEE/RSJ International Conference on Intelligent Robots and Systems, Sendai, Japan, pp. 2149–2154, September 2004
19. Andersen, T.T.: Optimizing the universal robots ROS driver. Technical report, Technical University of Denmark, Department of Electrical Engineering (2015)
20. Bradski, G.: The OpenCV library. Dr. Dobb's J. Softw. Tools **25**, 120–125 (2000)

Imagine the Future with Social Robots - World Robot Summit's Approach: Preliminary Investigation

Amy Eguchi[1(✉)] and Hiroyuki Okada[2]

[1] University of California, San Diego, La Jolla, CA, USA
a2eguchi@ucsd.edu
[2] Tamagawa University, Tamagawa-Gakuen, Tokyo 194-8610, Japan

Abstract. The One Hundred Year Study on Artificial Intelligence's Report of the 2015 Study Panel [1] predicts artificial intelligence (AI) to "enhance education at all levels, especially by providing personalization at scale" (p. 31). Interactive machines are already tutoring students in classrooms. It is anticipated that in the near future, social robots will become integral part of schools to enhance student learning experiences. This paper reports the preliminary investigation of participating students' experience through a new robotics competition focusing on the use of a social robot as a standard platform robot for primary and secondary school students. This report is part of a larger study of the World Robot Summit Junior Category'sSchool Robot Challenge. The School Robot Challenge at the Junior Category of the World Robot Summit (WRS), inspired by the WRS's aim to envision a society where humans and robots coexist and collaborate together, prepares students to participate in robotics and A.I. research and development in the future by offering a new robotics competition for students to design Co-Bot experience (human-robot co-existence) at school, hosted by the Japan Ministry of Economy, Trade, and Industry (METI) and the New Energy and Industrial Technology Development Organization (NEDO).

Keywords: Social robot · Human robot interaction · Educational robotics · Robotics competition

1 Introduction

Recent years, more social robots are becoming accessible for regular consumers to use in everyday life. At the CES 2018, following Jibo introduced in 2017, a handful of social robots including CLOi, BUDDY, and Tapia among many others were introduced. We categorized them as social home robots or partner robots designed to interact and communicate with humans. Sony's *aibo* was unleashed in November 2017, breaking more than 10 years of silence [2]. The new *aibo* robot features advanced mechatronics and AI to realize smarter and more sophisticated and natural performances resembling dogs. We are living in a transitional period when human-robot co-existence (Co-Bot) is becoming a norm in our society, similar to how personal computers and smartphones became essential parts of our lives. It is happening whether we are ready or not.

M. Moro et al. (Eds.): Edurobotics 2018, AISC 946, pp. 120–131, 2020.
https://doi.org/10.1007/978-3-030-18141-3_10

There is a growing need for Child-Robot Interaction (cHRI) research. There are a large number of opportunities where cHRI can be applied and make instantaneous impacts in education [3]. Social robots have started to be integrated into schools as part of educational/learning technologies. The roles that social robots have played in educational settings range from tutors [4, 5], teaching assistants [6], and fellow learners [7], to learning [8, 9] and therapeutic companions [10]. For example, NAO, a humanoid robot developed by Aldebaran Robotics has been used as a therapeutic robot with students with autistic spectrum disorder (i.e. [11, 12]). The main focus of current studies on the use of cHRI in education is on children as *consumers* or passive users of robots. Our study focuses on children as *active users* or *super users (producers)* rather than *consumers* of social robots in educational settings. With cHRI in educational settings, children are the center of the attention. The focus of all agents involved, including teacher(s) and robot(s), should be on maximizing student learning. Thus, we aim to encourage future generations to become *active* and *super users* of social robots by providing opportunities to explore ideas for how to best optimize the Co-Bot environment in school, and to engage students in programming a social robot to demonstrate their ideas through their participation in a robotics competition.

2 Method

This pilot study consists of multiple focuses for preliminary investigation. This paper is concerned with the investigation of students' experience programming a social robot to be used in an environment that they are familiar with – in school (Fig. 1), as demonstrated at the World Robot Summit (WRS) Junior Category, School Robot Challenge Workshop and Trial 2017 held in Tokyo, Japan in August 2017 [13, 14].

Fig. 1. Students programming and preparing for the presentation (The parent or guardian of all participants signed the photo image and video release consentform which was submitted to the secretariat of the World Robot Summit.)

2.1 The Competition

The Junior category of the World Robot Summit (WRS) offers a new robotics competition for students focus on how robotics can augment participating students' everyday lives. The WRS, hosted by the Japan Ministry of Economy, Trade, and Industry (METI) and the

New Energy and Industrial Technology Development Organization (NEDO), dedicates to creating a society where humans and robots coexist and collaborate in a way in which robots and humans augment each other in various aspects of our lives though the research and development of robotics technologies [13, 14]. In addition, it aims to deepen the understanding of and stimulate positive discussions on robots in society, leading to concrete suggestions and ideas on the applications of robotics technologies in future society. The WRS is scheduled to be held in October 2018 in Tokyo and October 2020 in Aichi prefecture, Japan. The WRS consists of four categories: Service Robotics, Industry Robotics, Disaster Robotics and Junior Robotics categories.

The Junior Category Challenges are inspired by the WRS's aim to envision a society where humans and robots coexist and collaborate together [13, 14]. In addition, it prepares students to participate in robotics and A.I. research and development in the future. There are two challenges in the Junior category – Junior Home Robot Challenge and Junior School Robot Challenge. Following the WRS's theme – *Robotics for Happiness*, the Junior Category Challenges ask students to create a scenario to achieve human-robot coexistence and cooperation at school or home. With the Junior Home Robot Challenge, a team of students develop a creative and innovative idea for the use of robots at home, and then realize their idea by programming a robot that they construct. With the Junior School Robot Challenge, a team of students are tasked to demonstrate their ideas for living and cooperating with a social robot in school. Each team develops a creative idea for using a social robot in school or at home, and realizes their ideas using social robots. The social robot used in the Junior School Robot Challenge is a standard platform.

2.2 Social Robot

The use of standard platform robot in competitions is not rare; however, the use of a sophisticated social robot with school-age children as a standard platform is quite unique [13, 14]. For example, VEX IQ Challenge is a competition for elementary and middle school students requiring teams to use only VEX IQ kits (https://www.vexrobotics.com/vexiq/competition). First LEGO League is a robotics competition for students (grades 4–8) using LEGO Mindstorms kits as the standard platform (https://www.firstinspires.org/robotics/fll). The social robot used in the WRS is the Pepper robot offered by SoftBank Robotics (Fig. 2). Pepper is currently available in a very few countries, including Japan, France, and Taiwan. Pepper is "the world's first personal robot that reads emotions" (http://www.softbank.jp/en/robot/) developed for domestic use.

Fig. 2. Pepper

Pepper is a 1.20-m-high humanoid robot that moves on three multi-dimensional wheels. It has two arms that can express its emotions through their movements [13, 14]. It is equipped with various sensors including a 3-D depth sensor, gyroscope, touch sensors, sonars, lasers as well as HD cameras. It has a touch screen on its chest with which users can control the robot using various apps. Pepper can be programmed with Choregraphe, a visual programming language, and other programming languages including Python, C++, and Java. It can be also controlled by Android and SoftBank Robotics provides SDK for Android Studio.

2.3 The Workshop and Trial

The WRS Junior Category - School Robot Challenge Workshop and Trial 2017 was held from August 2nd–6th, 2017 in Tokyo, Japan [13, 14]. There were 13 teams participating in the Workshop and Trial. Seven teams were from Japan (with one from a German School in Japan). Six teams were from Thailand, Vietnam, Malaysia, Australia, the Netherlands, and the U.S. The members of three Japanese teams were all female, and there were two mixed gender non-Japanese teams.

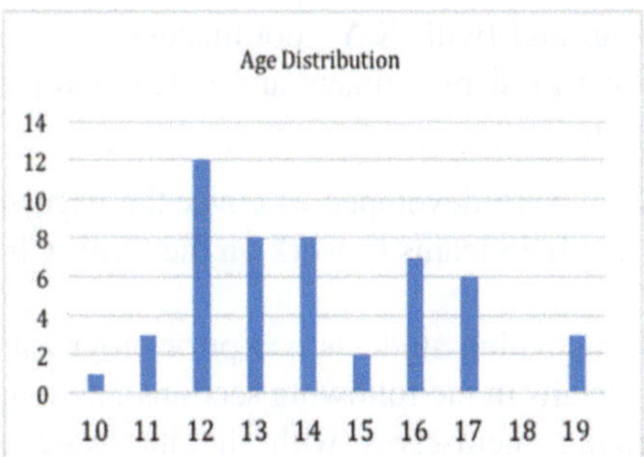

Fig. 3. Age distribution of the participants

There were 50 students participating. Among those, 28 were male (56%) and 22 were female (44%) students. In addition, 30 students (60%) were in middle school, while 16 high school (32%) and 4 elementary school (8%) students (Fig. 3).

The School Robot Challenge Workshop was held for two and half days focusing on programming for participating students [13, 14]. The goal of the Workshop was to provide opportunities for the participants to learn to program Pepper with Choregraphe. After the Workshop, teams presented their solutions to the Skills Challenges and Open Demonstration. Prior to the workshop, team mentors were provided with four documents (length in 20–30 pages) describing how to program Pepper using Choregraphe. However, since the robots could not be distributed outside of Japan, most of the teams did not have much experience programming Pepper before participating in the Workshop. The Workshop was held at Tamagawa Academy using their classrooms. There were three classrooms with an instructor in each classroom. Among three, two instructors provided the workshop in English, while one instructor provided the workshop in Japanese, mainly to Japanese teams. All Japanese teams had prior experience using Pepper; however, their experience programming Pepper was very minimum. The instructors followed the same workshop schedule and used the same workshop materials.

The Workshop started mid-day on August 2nd. Following two days, the teams gathered from 9:30 am to 4:40 pm to learn how to program Pepper [13, 14]. The Workshop, first, introduced the teams to Pepper's hardware focusing on how to handle the sophisticated social robot correctly, followed by the introduction to Choregraphe with the software interface and the important functions, including how to wirelessly connect their laptop with Pepper.

The programming lessons were structured as follows:

1. Make Pepper Speak
2. Make Pepper move its hands– using timeline-animation and timeline-parameter
3. Run multiple programs parallel
4. Use face-tracker function
5. Use speech recognition
6. Create a program to carry out a conversation
7. Use touch-sensors
8. Make Pepper move around (with X-Y coordinates)
9. Use tablet on the chest to display image and video, and play music
10. Use object recognition

The Workshop lessons were developed to cover the necessary functions and programming components that help teams to work on the Skills Challenges, as well as the Open Demonstration.

Since most of the participating students had prior programming experience, either with or without a robot (details in the following sections), although it started slow when students were familiarizing themselves with the hardware and the programming environment, soon they picked up the speed. Programming a social robot was also new to most of the team mentors (school teachers); however, their support during the Workshop also helped speeding up the students' learning. By the second day, older students were advancing the lessons following the lesson handouts without instructions by a instructor. This helped the instructors to focus on younger students who needed more hands-on supports. While the older and advanced students continued to learn to program Pepper with their own pace, the younger and less experienced students could also learn at their own pace with the instructor's customized instruction to meet the learning needs of the students.

On the third day, the teams mainly focused on creating solutions for the Skills Challenge tasks introduced on the first day and for their Open Demonstration task. The Skills Challenge tasks were:

1. Hold a scripted conversation (Fig. 4)
2. Identify and greet a person

3. Instruct a human partner to touch one of sensors. Once the human partner successfully touched the sensor, the robot gives a feedback that indicate that the task was successfully completed. Repeat the task with eight different sensors randomly
4. Identify a number on a 30 cm × 30 cm dice. The robot displays and call out the number to a human partner (Fig. 5)

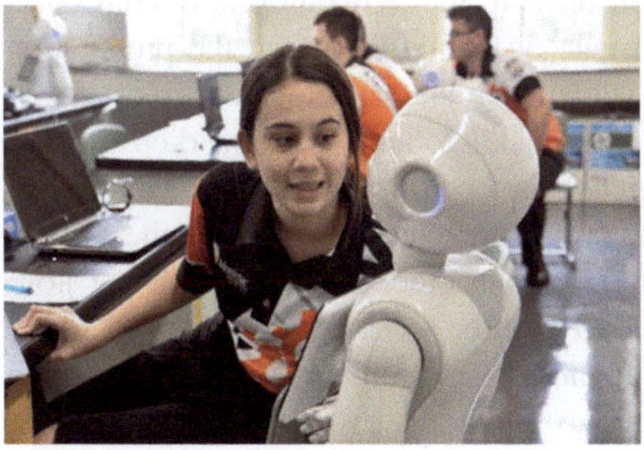

Fig. 4. Student working on the conversation task

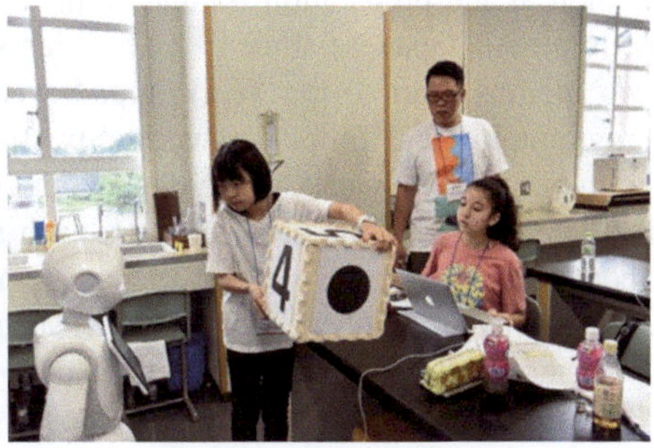

Fig. 5. The dice task

3 The Study

We conducted Pre- and post-questionnaires to collect data on the demographic information, preparation experience, perception of social robots, and STEM learning of the participating students. The questionnaires were sent to the participating students via their team mentor/contact adult. Among 50 participating students, 23 students (male, 12: female, 11) completed both pre- and post-questionnaires. Among those, 14 students were Japanese. Others were students from Australia (3; one female), Thailand (3; two females) and Malaysia (3).

3.1 Programming Experience

The data from the questionnaires shows that almost all participating students (96%) have some kind of prior programming experience. Among the students completed the questionnaire, 18 students (male, 8; female, 10) have prior experience programming robots. Three students (one female) have experience using Arduino with C++, while 14 students (male 6; female 8) used LEGO Mindstorms. Among those 14 students, one student has experience programming Mindstorms also with ROBOTC and Raspberry Pi with Python. Three students also have experience programming Arduino with C++.

Out of five students (one female) who did not have prior experience programming robots, four students (one female) have programming experience in other programming languages such as Java, JavaScript, SQL, Visual Basic.Net or C++.

All of non-Japanese team members (9; female, 3) have prior programming experience in text coding, while only one Japanese student (male) has experience with text coding.

3.2 Preparation

Out of 23 students, 20 students (male, 12; female, 8) started to learn Choregraphe or Python as the preparation for the competition prior to the Workshop and Trial competition. All Japanese students (12 students; male 6; female 6) started the preparation two months prior to the Workshop. Seven students (male, 3; female, 4) worked on programming 2–3 times per week. Six students (male, 2; female, 4) worked on programming once a week, while one student (male) worked on it every day. Majority of the students spent one to two hours each session (10; female, 7), while others spent about four hours each session (3, one female). On the other hand, most of the non-Japanese team members (6; female, 2) spent two to three weeks preparing prior to the Workshop and competition. All of the non-Japanese team members spent one to three days a week working on programming for one to two hours each session.

All Japanese team members (14 students) had a prior access to Pepper since they already had Pepper at their schools, while only three students from Australia had a prior access to Pepper. Australian team went to a local SoftBank facility to try out the hardware. They spent four hours at the facility programming Pepper using Choregraphe. The Japanese teams spent a couple of months with Pepper either by playing with it and/or programming it. The teams from Thailand and Malaysia (six students) had no experience with Pepper prior to the Workshop. However, two Malaysian students and one Thai student had programmed NAO with Choregraphe before. This means most of the students (87%) had prior programming experience with Pepper.

4 Student Experience

All 13 teams managed to develop their solutions for the Skills Challenge tasks and Open Demonstration by the end of the third day of the Workshop. However, there were several obstacles that students had to overcome to complete their demonstrations.

4.1 Technical Glitches

During the Workshop and Trial, there were some technical glitches caused by some uncertainty of organizing an event with a large number of robots all trying to communicate wirelessly with computers. During the Workshop, we set up one Wi-Fi spot in each classroom so that all Pepper robots could connect to laptop through the assigned Wi-Fi connection. Once an IP address was assigned to Pepper, students used the IP address to connect to their Pepper. Two Peppers were allocated to each team. There were 8–10 Peppers in each classroom during the Workshop (Fig. 6).

Fig. 6. Students during the workshop

Even with this setting, there were problems with Pepper's connection with a laptop. Since each student was working on a laptop while sharing two Peppers with other team members, they switched laptop connections with their robots quite often. If a laptop was steadily connected to one Pepper, the number of connection problems could have been reduced. When the connection problems persisted, students needed to restart the robot to reset the connection, which might have created another issue. Once a Pepper is restarted, usually a new IP address is assigned. However, since Choregraphe stores the previous connection, if something happens to a new connection, it could look for the previous IP address to try to connect to a robot. The IP address could have already been used by another robot, which would cause another connection problem. Because teams and instructors had to spend so much time solving the connection problems during the Workshop, the time for teams to actually work on creating their solutions to the Skills Challenge tasks and the Open Demonstration was significantly reduced, which caused some frustration among the teams.

The Trial competition was held in an open space which included a team-set up area. A stage was set up in the lower area of the same open space where teams demonstrated their solutions (Fig. 7). There was one Wi-Fi connection at the stage area and three other connections were provided at the team set-up area. Each team was instructed to use a specific Wi-Fi connection to avoid connection problems. However, the same problems we faced during the Workshop started to happen during the Trial. Peppers could access to any of the three Wi-Fi connections since it was an open space, which made it more challenging.

Fig. 7. Stage

Once multiple connection problems occurred with the assigned Wi-Fi connection, teams started to switch to another Wi-Fi connection creating another problem. When a connection trouble happens, Pepper would look for previously used Wi-Fi connection, which made it difficult to control the number of robots connected to one Wi-Fi connection. In addition, since there were many confusions among the teams, some mentors started to create hotspots using their own device. This created another condition that the organizers had difficulty controlling. Some teams ended up using an ethernet cable to connect to their robot to avoid the connection problems.

4.2 Lightings

The lighting on the stage was another problem that some teams faced. Although the organizers used shades to avoid natural lights coming through the windows, the spot lights on the stage were giving too much lights into Pepper's cameras. When Pepper is facing the audience, the lights would not be an issue. But when Pepper faced to the back wall on the stage, Pepper could not *see* an object. For example, when a student stood on the side of the stage wall and Pepper was facing the student, Pepper could not identify an object. When the student showed an object to Pepper, Pepper had to look up to see the object. That means that a direct light from the spot light could hit Pepper's cameras. When there is a direct light coming into the eyes, a human cannot see an object clearly. The same thing happens to Pepper. Many students were not aware of the effect on Pepper from the spot lights and struggled to make Pepper identify an object while demonstrating their solutions. Although it was an easy fix, it created some frustrations among the teams during the competition.

4.3 Student Performance

Although teams faced various obstacles, all 13 teams managed to demonstrate their solutions for the Skills Challenge tasks and Open Demonstration (Fig. 8). As the results from pre-/post-questionnaire show, most of the Japanese teams had some advantage during the preparation period including spending more time preparing for the competition while having an access to Pepper. However, the results suggest that their pre-competition experience did not give them observable advantages. Rather, what contributed to the ability to program Pepper to perform advanced tasks and/or complex performances was the text-coding experience of the team members. The team members who had programming experience with Python, whether with or without robot, could create complex codes with Python inside Choregraphe using Python boxes. Also, their understanding of various programming structures (i.e. nested loops) contributed to their ability to create complex codes with Choregraphe.

Fig. 8. Students demonstrating on stage

The observation data show that there were multiple factors that contributed to the success of the robot performance. Although team members' experience with text-coding seems to be a factor contributing to their ability to program with Choregraphe, another factor contributed to the successful on-stage performance seems to be the ability to develop a clean code. The teams that performed successfully on stage were both the teams with text-coding experience and the teams with clean coding structure, even it was less complex when compared with those created by the students with text-coding experience. Another interesting observation from the Trial competition was that high school teams created those complex codes, while the cleaner codes were developed by team members from elementary and middle school teams.

5 Perception of Social Robot

In the pre-/post-questionnaires, there were 14 questions related to the perception of social robot including social robots as friendly, kind, fun, a source of fear, intelligent, trustworthy, and helpful. The questionnaire used 5-point likert scale (5 = strongly agree and 1 = strongly disagree). In order to determine changes in students' perception of social robot as a result of their participation in the Workshop and Trial competition, pre- and post- comparison using two-tailed T-test were used. The preliminary results show that the majority of the items did not show significant changes, except for two items.

The first item that showed the change was the student's responses to "Humanoid robots will take jobs away from human". Students' response in post-questionnaire ($M = 3.30$, $SD = 1.18$) were significantly higher than those in pre-questionnaire ($M = 2.74$, $SD = 1.25$; $t(23) = 2.26$, $p < 0.03$). The result shows statistically significant increase in the student's perception of social robot to become a threat to future job. This could be because, through their firsthand interaction with Pepper, students might have learned the capabilities of current robotics technology and future possibilities.

Another item that showed the change was the student's responses to "Humanoid robot is fun to be with." Students' response in post-questionnaire ($M = 3.22$, $SD = 1.24$) were significantly lower than those in pre-questionnaire ($M = 3.87$, $SD = 1.25$; $t(23) = 2.92$, $p < 0.008$). The result shows very statistically significant decrease in the student's perception of social robot as an enjoyable companion. Their perception could have been influenced by their experience using Pepper through the Workshop and the Trial competition. As previously stated, almost all the teams experienced some degree of difficulties while working with Pepper such as communication problems and vision problems because of different lighting conditions. Their negative experience could have contributed to the result.

6 Ideas

Participating teams presented various ideas for social robots' role in school. Ideas from six teams related directly to student learning, and ideas from eight teams focused on robots helping school personnel. Among the teams presented the idea to create a robot to support student learning, three teams presented the ideas for study companion robots. Out of the three, one team also suggested that the study companion should act as a robot therapist. Three teams presented ideas to create a robotic teacher assistant in classroom. The teams that presented the robotic teacher assistant idea focused on making learning experiences at school *fun* by providing game-based learning or activities, as well as making learning at school *exciting* and *interactive* by having the robot provide information as on-demand learning. The concept of robots showing sympathy toward students was included in two teams' ideas for study companion robots, arguing that robots should provide comfort and show sympathy before helping students with their learning. One team presented the idea for Pepper to read student's facial expressions and check to see if s/he is okay before engaging in learning. Another team presented Pepper having the ability to cheer up a student who did not receive a satisfactory grade and encourage him/her to do better next time by studying together.

Ideas for robotic school personnel included a school guide (5 teams), a librarian's assistant, and a school nurse's assistant. The students felt that either there were not enough staff members to provide the desired support for students, teachers and guests at school or that the human personnel might be too old to extend his/her service to satisfy students' needs. It is evident that the teams were considering how robots could be designed to make lives at school better, not only for themselves but also for other people on campus.

7 Conclusion and Next Step

The purpose of this pilot study is to conduct a preliminary investigation to examine the capability of participating students in terms of programming a social robot as a standard platform, the strategies for using a social robot as a standard platform for a competition involving school-age students, the impacts that the experience programming a social robot could have on participating students' perception of social robots, and the students' ideas on social robots in school.

As described in the paper, the analysis of the data gathered from our observation and pre-/post-questionnaires highlight, although there are many issues need to be solved, the participating students were capable of programming Pepper with or without prior access to the hardware. However, their prior programming experience, with or without robots, could contribute to their successful learning of how to program Pepper.

Although the data is too preliminary to draw any conclusions, the experience working with social robot could have some influence on students' perceptions of social robots.

The ideas of social robot's role in school presented by participating teams at the Trial competition show that students care very much about the lives of people on campus, including school guards, librarians and nurses, as well as teachers and students. Their proposals for how using robots could *make life at school better* were very inclusive and aligned with the WRS's theme – *Robotics for Happiness.*

We plan to continue the pilot study with participating teams at the WRS 2018 competition in October 2018. Using the data from the two pilot studies, final data collection method at the WRS 2020 competition will be determined.

References

1. Study Panel - One Hundred Year Study on Artificial Intelligence, Artificial Intelligence and Life in 2030. Stanford University (2016)
2. Boyd, J.: Sony unleashes new aibo robot dog. In: IEEE Spectrum. IEEE (2017)
3. Belpaeme, T., et al.: Child-robot interaction: perspectives and challenges. In: ICSR 2013. LNCS. Springer (2013)
4. Vogt, P., et al.: Child-robot interactions for second language tutoring to preschool children. Front. Hum. Neurosci. **11**(73), 1–7 (2017)
5. Kennedy, J., et al.: Heart vs hard drive: children learn more from a human tutor than a social robot. In: The 11th ACM/IEEE International Conference on Human-Robot Interaction (HRI 2016). IEEE, Christchurch, New Zealand (2016)
6. Oh, K., Kim, M.: Social attributes of robotic products: observations of child-robot interactions in a school environment. Int. J. Des. **4**(1), 45–55 (2010)
7. Hood, D., Lemaignan, S., Dillenbourg, P.: When children teach a robot to write: an autonomous teachable humanoid which uses simulated handwriting. In: The Tenth Annual ACM/IEEE International Conference on Human-Robot Interaction. ACM, Portland, Oregon (2015)
8. Diyas, Y., et al.: Evaluating peer versus teacher robot within educational scenario of programming learning. In: The Eleventh ACM/IEEE International Conference on Human Robot Interaction. IEEE Press, Christchurch, New Zealand (2016)
9. Wong, C.J., et al.: Human-robot partnership: a study on collaborative storytelling. In: The Eleventh ACM/IEEE International Conference on Human Robot Interaction. IEEE Press, Christchurch, New Zealand (2016)
10. Arnold, L.: Emobie™: a robot companion for children with anxiety. In: The Eleventh ACM/IEEE International Conference on Human Robot Interaction. IEEE Press, Christchurch, New Zealand (2016)
11. Shamsuddin, S., et al.: Initial response of autistic children in human-robot interaction therapy with humanoid robot NAO. In: IEEE 8th International Colloquium on Signal Processing and Its Applications, Melaka (2012)
12. Tapus, A.A., et al.: Children with autism social engagement in interaction with Nao, an imitative robot: a series of single case experiments. Interact. Stud. **13**(3), 315–347 (2012)
13. Eguchi, A., Okada, H.: If you give students a social robot? - world robot summit pilot study. In: HRI 2018 Companion of the 2018 ACM/IEEE International Conference on Human-Robot Interaction. ACM, Chicago (2018)
14. Eguchi, A., Okada, H.: Learning with social robots — The World Robot Summit's approach. In: 2018 IEEE Integrated STEM Education Conference (ISEC). IEEE, Princeton (2018)

Technologies for Educational Robotics

Interdisciplinary Teaching with the Versatile Low-Cost Modular Robotic Platform EDMO

Rico Möckel(✉), Lucas Dahl, and Seethu M. Christopher

DKE SwarmLab, Department of Data Science and Knowledge Engineering, Maastricht University, Maastricht, The Netherlands
rico.mockel@maastrichtuniversity.nl

Abstract. The field of robotics embraces the interconnection of a variety of disciplines not only in research but also in teaching. For students this interconnection of disciplines can provide a different quality of understanding of theory and practice since it allows breaching, recapitulating, and interlinking knowledge and skills that otherwise might be taught only separately in a variety of well-contained courses. This paper describes the approach and material of a successfully running robotics course at Maastricht University that provides students with the opportunity to study and develop locomotion control with central pattern generators on a custom-made, low-cost, versatile EDucational MOdular robotic platform called EDMO. For these studies, students follow the chain of research and development from neuroscience over mathematical modeling, control theory, programming embedded systems, to the experimentation with robotic hardware that is supported by approaches from optimization and machine learning for optimal control parameter identification. The course thus interconnects a variety of disciplines and provides students with insights for instance into locomotion control and learning and provides a vivid view on aspects of numerical mathematics and calculus. We share teaching material and hardware design files for this course that is highly appreciated by our students in the hope that other teachers and students can benefit.

Keywords: Connectivism · Robotics in education · Active learning · Collaborative learning · Project-based learning · Modular robots · Central pattern generators

1 Introduction

Even though the idea of connectivism as a new independent theory of learning is not widely accepted [1,2], with its focus on building knowledge from networks of specialized information sets, the theory of connectivism [3,4] describes an important aspect that is widely accepted and experienced by many lecturers: students tend to obtain a different quality of understanding when they are presented

M. Moro et al. (Eds.): Edurobotics 2018, AISC 946, pp. 135–146, 2020.
https://doi.org/10.1007/978-3-030-18141-3_11

with the opportunity of connecting their knowledge obtained in different fields. In contrast to this observation, due to practical reasons, schools and universities often have to organize undergraduate teaching into separate courses that teach material according to their underlying separate fields and disciplines. To allow students to connect this separated material and knowledge, the Department of Data Science and Knowledge Engineering of Maastricht University decided to introduce project courses and meta-courses in its study programs that allow lecturers and students to bring together and revisit skills and knowledge that for organizational reasons had first been introduced in separate courses.

This paper describes the approach and content of one of the particularly successful meta-courses "Robotics and Embedded Systems" from the Bachelor study program "Data Science and Knowledge Engineering" - a course that combines elements from connectivism, active learning, and collaborative learning (see course overview in Sect. 2). Representing itself already as a highly interdisciplinary field that merges components from mechanical engineering, electrical engineering, computer science, mathematics, and control theory and that involves processes like design, construction, programming, and experimentation, robotics is well-suited as the central theme of such a meta-course.

In the presented course, the taught control approach of central pattern generators (CPG, described in Sect. 4) is inspired by the neural circuitry found in the spinal cord of vertebrates [5]. So in addition to the regular robotic disciplines, the presented course further combines aspects of biology, computational neuroscience, and mathematical modeling: As part of the course, students are challenged with the implementation and application of a CPG for locomotion control on a custom-made EDucational MOdular robotic platform called EDMO. Here the CPG is modeled as a set of coupled differential equations that mimic the dynamics of the mean-time firing rate of a network of neurons. To achieve efficient locomotion, the CPG parameters representing the coupling strength and firing parameters of neurons have to be tuned - a task that allows students to explore a variety of algorithms from machine learning and optimization (described in Sect. 4).

We chose to let students test their control on real-world hardware - rather than pure simulations - because like other lecturers in robotics, we experience the added value for students, when allowing them to explore, to understand, and to learn to exploit and compensate for real-world imperfections and noise. Since locomotion requires tight interaction between the robotic hardware and its environment, the task of locomotion control is especially well-suited for exploring robotic hardware in real-world environments. The EDMO platform (described in Sect. 3) was designed as a versatile, modular robotic platform that allows students to construct and explore a variety of robotic morphologies including snake, biped, and quadruped configurations. Several example configurations of EDMO modules are depicted in Fig. 1. EDMO modules have been designed with a philosophy in mind that explicitly allows students to make and to actively learn from mistakes (Sect. 3). EDMO modules combine low-cost off-the-shelf servo motors with low-cost custom-made 3D printed shells. To drive the EDMO

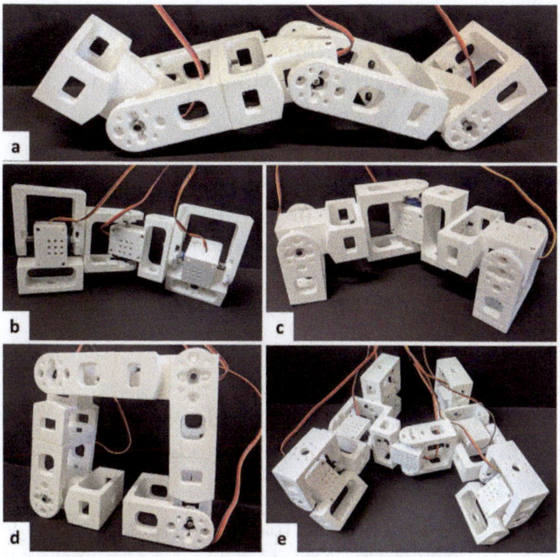

Fig. 1. Examples of robot configurations composed of EDMO modules: (a) Three sequentially connected EDMO modules form a snake robot. (b) Biped configuration with two limb modules and one spine module. (c) Crab configuration. (d) Puppy configuration with four sequentially connected EDMO modules. (e) Quadruped robot with one spine-, two hip-, and four limb modules

robots, low-cost off-the-shelf electronics provided by Arduino are being used. EDMO modules are easy to reproduce, repair, and modify.

Emphasis has been given to make the course scalable and the required investments for teaching sustainable. Scalability is important since we and colleagues from other universities experience the continuously increasing demand of university education in engineering, computer science, and robotics. The number of students in the presented course has doubled over the past three years and we expect an ongoing growth in student numbers in the future. This continuous increase in student numbers means that we had to find a way to easily reproduce hardware setups. To make these continuous investments sustainable we chose technologies and off-the-shelf components that are likely to remain available in the coming years so that we can keep building and purchasing further hardware setups - rather than being faced with the situation that all setups have to be replaced because components are discontinued.

The use and combination of EDMO modules does not require any prior knowledge in mechanical engineering - making them also suitable for education with students from non-engineering disciplines and children. The simplicity of use was chosen so that if required, the robotics course could also be taught as an introduction to robotics course and to computer science students that have not received any prior university education in other engineering disciplines. By now we have tested the EDMO modules over two years with students during the presented course. In addition, we regularly use EDMO robots for public outreach and present the results of the student competition that concludes the course (see Sect. 5) during open days of Maastricht University, visits to schools, kindergartens, and homes for elderly.

The popularity of the course and the enthusiasm of children, students, teaching assistants, visitors, and lecturers when learning, experimenting, and teaching with the EDMO robots encouraged us to present this paper in the hope that other universities, schools, and private people can also benefit from these developments.

2 Course Overview

An overview of the course content is given in Table 1. The course can be roughly divided into four phases designed to support well-defined learning objectives.

Table 1. Overview of course content and related learning goals. L: lecture, D: demonstrations, T: hands-on tutorial with electronics and robotic hardware.

Phase	Content	Learning objectives
Continuous:	Discussion of state-of-the-art robotic systems and challenges	01, 02, 10
Phase 1: Tutorials & Lectures	LD: Introduction to robotics	01, 02
	LD: Embedded systems & Real-time	03, 06, 08
	LDT: Introduction to Arduino and C	03, 04, 05, 06
	DT: Interrupts & Timers	03, 04, 05, 06, 07, 08
	LDT: Analog-to-digital conversion	03, 04, 06, 07, 08
	LDT: Pulse-width modulation	03, 04, 06, 07, 10
	LDT: Communication	03, 04, 05, 06, 07, 08
Phase 2: Assignments 1 PID	LD: Actuation and PID control	01, 02, 03, 07, 08, 09 , 10
	6h group assignments: PID control	03, 04, 05, 06, 07, 08, 09, ... 10, 11, 12
Phase 3: Assignments 2 CPG	LD: Dynamic systems and CPG	01, 02, 03, 09
	6h group assignments: Locomotion control with CPG	03, 04, 05, 06, 07, 08, 09, ... 10, 11, 12
Phase 4: Competition & Outreach	Student competition	03, 04, 06, 08, 09, 10, 11
Final exam	Individual written closed-book exam	01, 02, 03, 04, 08, 09, 10

During phase 1, students follow lectures and hands-on tutorials with an Arduino platform. At Maastricht University, courses are offered in the form of 2 h sessions - often lectures and tutorials are divided into different sessions. In the presented course we changed the format so that lectures and tutorials are provided interleaving with short time intervals to keep students focused and well-motivated. As a result, during a 2 h session, there might be a 20 min lecture on the theory of real-time computing, followed by a 10 min tutorial on how to use hardware timers of the Arduino platform to control an LED, followed by 20 min where students try the newly obtained knowledge and skills themselves, followed by a 10 min post-discussion. Then the next topic might be introduced with a short lecture, demonstration, and tutorial.

In phase 2 and 3, students are introduced to and work on two practical assignments in groups of 2–3 students: In assignments 1 (phase 2), students build a servo motor from a brushed DC motor, potentiometer, microprocessor, and DC motor driver. Students implement and adjust the parameters of a PID controller to enable stable position control with the self-made servo motor. In assignments 2 (phase 3), students program and experiment with the parameters of a locomotion controller based on a central pattern generator (CPG) for controlling a self-made robot composed of EDMO robot modules.

The course is concluded by a student robot competition (phase 4) where students compete in two categories: (1) Building the fastest robot. (2) Building the most creative moving robot. Students obtain a final course grade based on their assignment results and answers during a final written exam.

With the course, we pursue a variety of learning objectives (LearnObj): **LearnObj01:** Students gain a broad understanding of what is a robotic system and the variety of existing robotic systems. **LearnObj02:** Students gain the ability to critically study and judge robotic systems from demonstrations. **LearnObj03:** Students learn to understand the connection between and the concepts of robotics, embedded systems, and real-time processing. **LearnObj04:** Students obtain a general understanding of the architecture of a microprocessor, its core, and its peripherals. Understand the difference between a general purpose CPU and a microprocessor. Understand the purpose of peripherals like analog-to-digital converters, digital-to-analog converters, pulse-width modulators, interrupts, UART, I2C, SPI, CAN, and USB communication interfaces. **LearnObj05:** Learn how to create more efficient code. Get acquainted with basic and advanced concepts of the programming language C on an Arduino platform. **LearnObj06:** Students learn how to deal with the lack of an operating system. **LearnObj07:** Students obtain basic understanding on how to read and use a circuit diagram. **LearnObj08:** Students obtain a fundamental intuition for the control of real-world systems. Learn how to deal with and to compensate for imperfections and noise. **LearnObj09:** Students practice knowledge from different mathematical disciplines and see mathematical systems in action. By applying PID and CPG control, students have the opportunity to experience the application of dynamic systems, ordinary differential equations, and numerical mathematics for controlling real-world systems. **LearnObj10:** Students gain a general understanding of actuation, motors, motor drivers, and motor control. **LearnObj11:** Students practice group work. **LearnObj12:** Students learn how to conduct and document an experiment with a robotic platform. Table 1 shows which learning goals are targeted in the different lectures, demonstrations, and tutorials that are preparing students for the practical assignments as well as the strong focus of achieved learning goals during the assignments.

3 EDMO - Design and Hardware

In comparison to monolithic robots, modular or cellular robots are composed of a set of homogeneous or heterogeneous modules that can be combined to form a variety of robot configurations (for an example see Fig. 1). Due to this ability to form many configurations and its modularity it is often argued that modular robots are well suited for tasks that require versatility and robustness including space exploration and disaster management (see [6] for a popular overview of modular robots): A damaged module can be easily replaced without sacrificing the rest of the robotic hardware. If a different task should be fulfilled, the structure of the robot can be changed by reconnecting, adding or removing modules - rather than designing a new monolithic robot from scratch.

These properties make modular robots more scalable than monolithic robots and thus also ideal for teaching where students should be allowed to break some modules, but need to be able to continue their studies quickly without the tedious task of identifying and repairing the damage in a complex robot as it is typically the case for monolithic robots. In the literature a large variety of studies can be found that showcase the versatility and scalability of modular robots. One of the authors of this paper for instance contributed to the two systems YaMoR [7] and Roombots [8,9]. In 2014 we identified more than 72 modular robotic systems [9]. Their number keeps growing. Most of the existing modular robots are designed to study specific research questions. But some publications also mention the purpose of the robots for education.

There are some key challenges however, when applying modular robots for large-scale education that many publications do not target too explicitly:

(1) *Costs:* When trying to design modules that are autonomous, costs of these individual modules is typically too high to afford building a large number of copies. As a result typically there exist only a few copies of those modular robots from the literature. (2) *Production and maintenance:* Making compact autonomous robotic modules is a challenge by itself - often requiring custom-made parts that are difficult to reproduce, expensive electronics, and skilled engineers for building and maintenance. (3) *Sustainability and scalability:* The dependence on custom-made parts that are difficult to manufacture, off-the-shelf components that are susceptible to be discontinued, and custom-made software that is likely to not to be well-documented and supported makes some robotic platforms unattractive for teaching where teachers are faced with courses where the numbers of students continuously keeps increasing, and there is a continuous need to maintain, repair, and replace robotic hardware with a limited teaching budget and personnel.

Table 2. Overview of EDMO hardware for assignment 2 - CPG controlled snake robot.

Component	Specification	Weight [g]	Price [€]
Plastic parts	3D printed body parts	51	<5.50
Servo	Turnigy^{TM}, TGY-50090M Analog servo (Torque: 1.6 kg·cm (15.7 Ncm) at 4.8 V, Speed: 0.08 sec/60° at 4.8 V)	15	<5.50
Accessories	Fasteners and cables	15	<1.50
One EDMO module		**81**	**<12.50**
3 EDMO modules			**<38.00**
Arduino nano	Microprocessor board		<7.00
Breadboard	Extension board for Arduino nano		<2.00
Power converter	5 V DC - 2 A Raspberry Pi power converter		<7.00
Accessories	Connectors and cables		<2.00
Total electronics			**<18.00**
Costs/Student group			**<56.00**

Table 2 provides an overview of the EDMO hardware components. EDMO modules have been designed to benefit from the advantages of modular robots while targeting the aforementioned challenges when applying modular robotic hardware in teaching: The open-source EDMO modules are composed of only three 3D printed parts and an off-the-shelf servo motor. Figure 2 shows both the full EDMO module as well as its disassembled individual components. The connection between 3D printed parts and a servo motor with off-the shelf screws and nuts requires little skill and no special tools - making assembly and maintenance a fast and low-cost process. New modules can be assembled in less than 3 min. 3D printing technology and the choice of a low-cost servo motor that comes with a standard form factor supported by several motor brands guarantee that teachers and students can keep producing additional modules in the future whenever needed. The 3D printed parts are very robust and unlikely to break - even through high-impact gaits. During two years of teaching and intensive testing, we never experienced a broken 3D printed part. The parts were designed so that a fabrication with a variety of low-cost 3D printing processes is supported, including acrylonitrile butadiene styrene (ABS) or polylactic acid (PLA). Printing is possible in professional workshops for low-cost, high-volume production or for small volume production with small size hobby printers (building volume larger than 690 mm x 670 mm x 300 mm) using only a single extruder and without any support material. To achieve low weights and to save printing material, 3D parts are printed using a light cross print pattern with a low to medium print strength for the internal structures. To form larger configurations, EDMO modules are connected with each other through off-the-shelf screws.

EDMO modules were designed with the idea in mind that students must have the opportunity to learn without fear of breaking the hardware. When studying locomotion, high impact gaits cannot always be avoided. When studying the

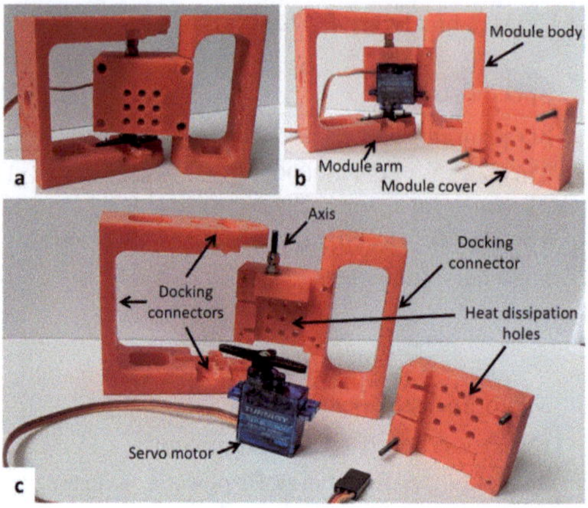

Fig. 2. (a) A single fully assembled EDMO module. (b) The three main mechanical components of an EDMO module, depicted with removed cover. (c) Disassembled EDMO module.

control of complex robot configurations, students can generate kinematic chains with self-collisions. To reduce the risk of breaking parts, we could have designed with soft robotic parts that are difficult to control, or in a very robust manner that most likely would have led to heavy, expensive, and potentially dangerous robotic modules. Instead, EDMO modules have been designed with a single, well-controlled breaking point - the gearbox of the off-the-shelf servo motor - making it easy and cheap to repair EDMO modules by the students themselves - thus unloading teachers and teaching assistants from the maintenance of the modules. In the past two years we experienced that on average one out of 10 motors gets broken during a course, typically leading to a steep learning where students start to understand why they broke the motor and how to avoid a similar situation in the future.

We decided to drive EDMO modules by Arduino electronics because it is easily available, low-cost, widely supported, and provide easy to use integrated development environments at no costs. Arduino electronics are being used during all phases of the course so that students get slowly introduced to the concept of programming embedded electronics. Teachers have the choice between very low-cost, compact electronics like the Arduino nano with 8-bit processors in combination with a breadboard or more expensive Arduino computation boards with 16 or 32-bit processors. Different versions of the Arduino electronics support a variety of compatible extensions like pluggable motor driver boards that support the tasks of the practical group assignment 1 (PID control) of the course. An alternative low-cost motor driver is the popular L298N module that is available at less than 3€. For assignment 2 (Locomotion control with CPG) we supply a team of 3 students with 3 EDMO modules and one set of electronics leading to a total cost of less than 19€ per student as shown in Table 2.

4 Locomotion Control with Central Pattern Generators

Central pattern generators (CPG) are biological neural networks capable of producing rhythmic patterns without the need of any sensory or motor input. CPGs can be modeled in a biologically more plausible way with bidirectionally coupled clusters of inhibitory and excitatory spiking neurons and have been studied in combination with locomotion generation in animals and robots [5,10]. We and many others studied locomotion control and learning with CPGs and modular robots [7–9].

CPGs are interesting for teaching dynamic systems because they can also be modeled in a compact, biologically less plausible way in form of phase-coupled oscillators with a set of coupled differential equations (Eqs. 1–3) that can be implemented with a few lines of plain code and be executed efficiently on a low-cost microprocessor:

$$\dot{\phi}_i = 2\pi f + \sum_{i \neq j} w_{ij} r_i \sin(\phi_j - \phi_i - \psi_{ij}) \tag{1}$$

$$\dot{r}_i = a_i(R_i - r_i) \tag{2}$$

$$\Theta_i = r_i \sin(\phi_i) + X_i \tag{3}$$

where $\dot{\phi}_i$ denotes the phase velocity of an oscillator i, r_i the oscillator's amplitude, and Θ_i its output in rad that can be used to directly set the angle of a servo motor.

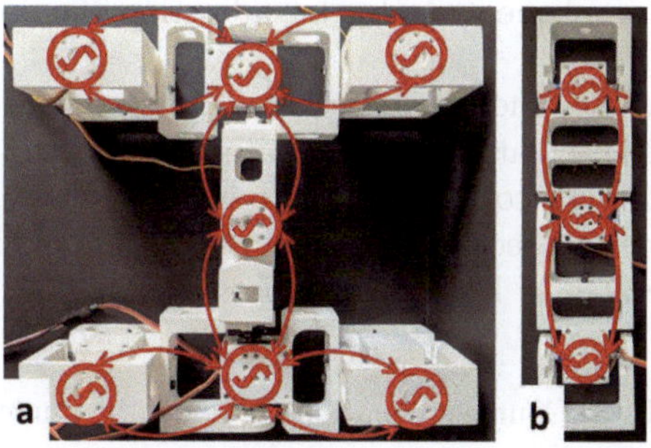

Fig. 3. EDMO (a) quadruped and (b) snake configuration with respective CPG networks of bidirectionally coupled oscillators.

When applying CPGs to locomotion control of robotic structures, typically a single oscillator per degree of freedom is sufficient. Figure 3 shows a quadruped and a snake configuration composed of EDMO modules with overlayed networks of oscillators. To achieve efficient locomotion patterns only three control parameters per oscillator have to be optimized: (1) The oscillators' desired offsets X_i, (2) the desired amplitude of each oscillator R_i, and (3) the coupling phase bias,

ψ_{ij}, that controls the desired phase difference between each pair of oscillators i and j. This number of parameters can be further reduced if symmetries of the robotic structure can be exploited (like in the quadruped in Fig. 3a) or parameters can be shared because oscillators are organized in form of a chain (like in the snake robot in Fig. 3b). The parameter f that controls the frequency of the oscillation in steady-state typically can be shared among all oscillators and does not have to be optimized but can be used as a parameter to adjust the speed of the robot after the optimization process is finished. A variety of popular optimization approaches including Evolutionary and Particle Swarm Optimization Algorithms can be taught and practiced for locomotion control and learning with central pattern generators. In the presented course students get encourage to apply Powell's method - a gradient-free optimization approach [11], that supports online learning of locomotion parameters and that in comparison to other popular optimization techniques in practice often leads to good locomotion results with a relative low number of hardware experiments [7].

The remaining parameters of the Eqs. 1–3 can be set a priory to fixed values: The parameter w_{ij} denotes the coupling strength of pairs of oscillators and controls how quickly Eq. 1 adjusts the corresponding phases of the oscillators i and j after the desired phase bias ψ_{ij} has been changed. Equation 1 achieves this adjustment of phases by increasing or decreasing the phase velocity $\dot{\phi}_i$ of the corresponding oscillators. Equation 2 implements proportional control for the oscillators' amplitudes r_i to achieve smooth adjustments of an oscillator's actual amplitude r_i after a discrete change of the desired amplitude R_i by adjusting the required change in amplitude $\dot{r}_i$. The rate of change is controlled by the parameter a_i that corresponds to a proportional control parameter. A similar mechanism could be implemented also for adjusting the oscillators' offsets but is not shown here.

Equation 3 finally uses the generated phase, amplitude, and offset variables to generate the desired oscillation. To calculate this output, the phase and amplitude of each oscillator have to be first determined from the given phase velocity and rate of amplitude, respectively, through a standard numerical integration method like Runge-Kutta or Euler integration.

Table 3. Course evaluations provided by students.

Academic year	# students who took course exam	# students who evaluated course	Course grade [Likert Scale]	Standard deviation (course grade)	Teacher grade [Likert Scale]	Standard deviation (teacher grade)
2015–2016	28	16	4.5	0.63	4.88	0.34
2016–2017	23	4	4.33	0.47	4.75	0.43
2017–2018	41	22	4.68	0.55	4.91	0.28
Mean	30.7	14	4.5	0.55	4.85	0.35

5 Course Evaluation and Conclusions

Students at Maastricht University are invited to evaluate courses after taking the exam. During these fully anonymous, voluntary evaluations, students grade the overall course, its components, teachers, and teaching assistants on a 5-point Likert Scale [12], where a grade of 5 corresponds to the highest level of student satisfaction. In addition, students can use the opportunity to provide comments and feedback in the form of plain text. The results of the student evaluations for the presented course are provided in Table 3.

Overall, we observed that in the past years on average only 46% of the students taking the course exam also provided an evaluation. At Maastricht University, students are not obliged to participate in the evaluation. From personal conversations with our students, we found that the more satisfying a course is, the less students are motivated to provide feedback because they do not see a reason for constructive feedback if a course is very satisfying. This caused us in the present academic year 2017–2018 to explicitly motivate students to participate in the evaluations by explaining the importance also of positive feedback.

Given that an overall average course grade of 5 is hardly ever provided by students, the overall course grades are very good and among the top grades given by students to a course in the study program of Data Science and Knowledge Engineering at Maastricht University. These results have to be seen also in relation to the fact that this DKE study program has been the highest-ranked Dutch Bachelor program in the field of Artificial Intelligence for the past 5 consecutive years (2014–2018) according to the influential Dutch University Ranking *Keuzegids Hoger Onderwijs*.

The high course grade is consistently motivated by students in the free text feedback part of the evaluation by: (1) Students appreciate the strong connection between theory, including algorithms from optimization, control theory, dynamic systems, and numerical mathematics on one side and the practical applications of motor and locomotion control and learning by programming embedded systems and experimenting with robotic hardware on the other side. Some students even claim that the course was the first time where they truly understood dynamic systems. (2) Students are very satisfied with the challenging practical assignments. Here the course benefits from the scalability of our teaching approach with EDMO modular robots and CPG control that allow us to adjust the level of difficulty to the specific level of skill and knowledge of the individual student groups within one course. Students who require more practice and time can optimize CPG parameters of a snake robot - composed of three EDMO modules - by hand while highly ambitious student groups can implement and run full optimization algorithms on complex robotic structures like quadruped robots - thereby exploring different gait patterns, steering capabilities of the robotic structure, and optimizing its robustness with regard to different terrains. (3) Students highly appreciate the final robot competition where they can design robotic structures from EDMO modules and try out control according to their own interest to build the fastest and most creative moving robot.

In the hope that our work on EDMO and the related course content will be beneficial for other teachers and students outside of our department, we provide continuously updated material online including access to all design files at https://project.dke.maastrichtuniversity.nl/SwarmLab/edmo/.

Acknowledgment. The lecturer of this course, Dr. Rico Möckel, would like to express his gratitude to his former mentors Prof. Auke Ijspeert from the École Polytechnique Fédérale de Lausanne (EPFL), Switzerland, and Prof. André Seyfarth from the Technische Universität Darmstadt (TU Darmstadt), Germany, for their commitment and support when sharing their teaching philosophies and experience with the author during his Postdocs under their supervision.

We thank the many students who participated in this course and who made this course very enjoyable with their enthusiasm, commitment, and creativity.

References

1. Verhagen, P.: Connectivism: a new learning theory? (2006). https://www.scribd.com/doc/88324962/Connectivism-a-New-Learning-Theory
2. Kop, R., Hill, A.: Connectivism: Learning theory of the future or vestige of the past? Int. Rev. Res. Open Distrib. Learn. **9**(3), 1–13 (2008). https://doi.org/10.19173/irrodl.v9i3.523
3. Siemens, G.: Connectivism: A learning theory for the digital age (2014). http://www.itdl.org/journal/jan_05/article01.htm
4. AlDahdouh, A., Osorio, A., Caires, S.: Understanding knowledge network, learning and connectivism. Int. J. Instr. Technol. Distance Learn. **12**(10), 3–21 (2015)
5. Ijspeert, A.J.: Central pattern generators for locomotion control in animals and robots: A review. Neural Netw. **21**(4), 642–653 (2008). https://doi.org/10.1016/j.neunet.2008.03.014
6. Yim, M., Shen, W.-M., Salemi, B., Rus, D., Moll, M., Lipson, H., Klavins, E., Chirikjian, G.S.: Modular self-reconfigurable robot systems [grand challenges of robotics]. IEEE Robot. Autom. Mag. **14**(1), 43–52 (2007). https://doi.org/10.1109/MRA.2007.339623
7. Sproewitz, A., Moeckel, R., Maye, J., Ijspeert, A.J.: Learning to move in modular robots using central pattern generators and online optimization. Int. J. Robot. Res. **27**(3–4), 423–443 (2008). https://doi.org/10.1177/0278364907088401
8. Sprowitz, A., Pouya, S., Bonardi, S., Van Den Kieboom, J., Mockel, R., Billard, A., Dillenbourg, P., Ijspeert, A.J.: Roombots: reconfigurable robots for adaptive furniture. IEEE Comput. Intell. Mag. **5**(3), 20–32 (2010). https://doi.org/10.1109/MCI.2010.937320
9. Spröwitz, A., Moeckel, R., Vespignani, M., Bonardi, S., Ijspeert, A.J.: Roombots: A hardware perspective on 3D self-reconfiguration and locomotion with a homogeneous modular robot. Robot. Auton. Syst. **62**(7), 1016–1033 (2014). https://doi.org/10.1016/j.robot.2013.08.011
10. Ijspeert, A.J., Crespi, A., Ryczko, D., Cabelguen, J.-M.: From swimming to walking with a salamander robot driven by a spinal cord model. Science **315**(5817), 1416–1420 (2007). https://doi.org/10.1126/science.1138353
11. Press, W.H., Teukolsky, S.A., Vetterling, W.T., Flannery, B.P.: Numerical Recipes: The Art of Scientific Computing, 3rd edn. Cambridge University Press, New York (2007)
12. Likert, R.: A technique for the measurement of attitudes. Arch. Psychol. **22**(140), 1–55 (1932)

Engaging Learners in Dialogue Interactivity Development for Mobile Robots

Paul Baxter(✉), Francesco Del Duchetto, and Marc Hanheide

Lincoln Centre for Autonomous Systems, School of Computer Science, University of Lincoln, Lincoln, UK
{pbaxter,fdelduchetto,mhanheide}@lincoln.ac.uk,
https://lcas.lincoln.ac.uk

Abstract. The use of robots in educational and STEM engagement activities is widespread. In this paper we describe a system developed for engaging learners with the design of dialogue-based interactivity for mobile robots. With an emphasis on a web-based solution that is grounded in both a real robot system and a real application domain – a museum guide robot – our intent is to enhance the benefits to both driving research through potential user-group engagement, and enhancing motivation by providing a real application context for the learners involved. The proposed system is designed to be highly scalable to both many simultaneous users and to users of different age groups, and specifically enables direct deployment of implemented systems onto both real and simulated robots. Our observations from preliminary events, involving both children and adults, support the view that the system is both usable and successful in supporting engagement with the dialogue interactivity problem presented to the participants, with indications that this engagement can persist over an extended period of time.

Keywords: Mobile robots · DialogFlow · Museum guide · Public engagement · Dialogue interactivity

1 Introduction

Robots have long formed a useful educational tool in part by making computation tangible and accessible [12], as physical instantiations of Papert's communication with computers [13]. We are concerned here with robots acting as educational platforms that the learners *control*, rather than the more recent endeavours to use *social* robots as learning companions, for which a different set of considerations apply, e.g. [1]. More specifically, we are interested in encouraging engagement with the high-level control of robots (rather than the low-level

The work described in this paper was supported by the EU H2020 ("ILIAD", contract 732737), EPSRC ("NCNR" Hub EP/R02572X/1), and the Johnson and Mukherjee Brothers Charitable Trust.

M. Moro et al. (Eds.): Edurobotics 2018, AISC 946, pp. 147–160, 2020.
https://doi.org/10.1007/978-3-030-18141-3_12

control of teleoperated movement for example) such as social interactivity, and the concepts involved in programming these into robots.

The work we present in this paper is part of a wider research programme concerned with improving the long-term autonomy of mobile robots in human-centred environments, as an extension of prior and continuing research [4,7]. Of specific interest to the current effort is to increase the social interactivity of the robots concerned so as to reduce barriers to interaction and increase utility for non-technical users. One application domain of particular interest is the use of a mobile robot to interact with the general public in a museum, to help guide, provide information, and engage members of the public with the exhibits. The general research context for the present paper is that one potentially valuable source of information on interactions with the public can be obtained through the public's engagement with robot-centred STEM activities.

This work is conducted in collaboration with the Collection Museum[1], which houses exhibits related to local archaeology and history. In an ongoing collaboration, the intention is for a robot to be permanently based in the museum, to act as a guide, and to provide further information on exhibits to interested visitors. In order to achieve this, the robot should engage members of the public who have no guaranteed prior experience with robots, and be able to respond to queries in an appropriate manner. While the robots are equipped with a touchscreen that may be used for interaction (Sect. 3.3), the intention is to enable verbal interactions, thus necessitating dialogue management, since verbal interactivity is an important means of guiding users' expectations of competence [6].

In this paper, we bring together these strands of educational support through engagement with robotics, and research efforts to increase social interactivity of autonomous robots. By integrating these efforts, we seek to enhance the benefits to both - in terms of driving research through engagement with potential user groups (towards the manner of participatory design [14] for example), and enhancing motivation by providing a real application context for the learners involved.

We introduce a system that supports this integrative effort: it provides a tool to engage with the complexities of human-like dialogue, and links this to both real and simulated robots so that these issues can be directly evaluated. Being a web-based application, this does not require the installation of specialist software (beyond a standard web browser), and enables the simultaneous participation of large numbers of users. As a result, the learners not only explore issues related to social interactivity in our application domain, but also gain hands-on experience with the latest technology used for dialogue interactivity with computers more generally. We intend this paper to act as a resource for those who wish to use and evaluate (beyond our case studies) the system we present: to facilitate this, all code sources and exercises are provided.

[1] https://www.thecollectionmuseum.com/.

2 Background

Robot programming in the educational contexts of interest here have typically focused on using robots to physically act out the implementation of algorithms (e.g. controlling movement). From the basic use of programming languages, tangible devices were introduced for such implementations [12]. More recently graphical programming languages (such as Scratch[2] or Blockly[3]) have been used to provide accessible means of learning programming principles, despite some evidence of superior engagement with tangible devices [8]. The intent in these is to enable non-experts to engage with and learn the basic concepts and control, particularly where robots are concerned.

A range of methods have been employed in terms of programming a robot to achieve desired tasks. In addition to explicit and graphical programming, where a high level of platform familiarity and programming expertise is required, there have been increasing efforts to have the robots learn from demonstration, e.g. [2]. An approach such as this more explicitly acknowledges the role that non-robotics-experts can play in increasing the competence of robots. This acknowledgement continues with programming methods that abstract away from low-level algorithms and focus on behaviours, e.g. for social robots [5].

A number of robots have in the past been used in the domain of museums for guiding visitors. A notable example is the Minerva robot [15], which allowed the general public to control the robot through a web interface, enabling both influence of high-level strategy (choosing a visitor tour), or by explicitly choosing a predefined location in the museum for the robot to navigate to. While the robot was capable of some social interaction and was able to learn during interactions, this could not be explicitly influenced by the public controllers. More recent applications have, for example, focused on adding more human-like guiding behaviours to robot tour guides, with the intention of improving the experience for museum attendees [9]. Results indicate that human-like communication is used when interacting with robots [10].

The challenge thus remains improving the social interactivity of the robots that help the visitors. As noted above, we focus on language-based interactivity, and how we can engage naive users, whether they are children or adults, with the fundamental issues involved and potential solutions.

3 System Setup

In order to improve the dialogue handling capability of our museum guide robot, we have designed a system that is comprised of three main elements (Fig. 1). First is DialogFlow, a tool (now under the Google umbrella) to design, implement, and host conversational interfaces (whether these are text or speech based). Second is a robot system, in our case operating on the Robot Operating System (ROS), which can either be real or simulated, to which implemented DialogFlow

[2] https://scratch.mit.edu/.
[3] https://developers.google.com/blockly/.

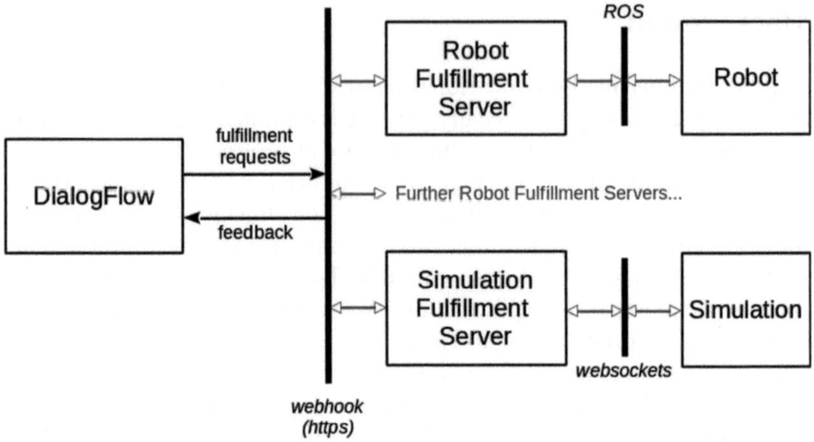

Fig. 1. System overview: the robot and the simulation can be swapped without behaviour reconfiguration by simply defining an alternative fulfillment server. The robot fulfillment server and robot communicate through ROS, and the simulation fulfillment server and the simulation through web sockets. The interaction between DialogFlow and the fulfillment server uses a webhook (https).

behaviours are linked. Finally is a fulfilment server that link these two together. In the following subsections, we briefly describe each of these components. The source code is freely available for inspection and use[4].

There are three main benefits to using a system such as this. Firstly, due to the availability of the simulated version of the robot and environment, the activities can be completed anywhere with a computer and an internet connection. This means it is versatile and applicable for the vast majority of engagement scenarios (from school visits to/from University, workshops at summer schools/camps, etc). Secondly, the online nature of the tools used means that the engagement from participants can continue beyond the formal activity time. This is an important facet of the current system, as it facilitates ongoing interaction and learning for the participants on their own terms. Finally, it enables robot behaviours developed in simulation, and potentially remotely, to be applied to a real robot in a real environment with minimal alteration. This is a useful feature in terms of applicability to the research context, but it can also be used to support motivation through the potential for the fast deployment of participant behaviours.

3.1 DialogFlow

Handling natural language is an important feature for a robot intended to interact with the public. While natural language processing encompasses a wide range of technical challenges [3], the aspect that we are concerned with particularly

[4] https://github.com/LCAS/robot-dialogflow/.

is dialogue management, which is concerned more specifically with linking the current state of the conversation with actions (which could include further dialogue) [11]. There are a wide range of approaches and tools available for doing this. For reasons of availability, usability, and scalability, we employ DialogFlow[5], a Google-owned tool that facilitates the design of appropriate responses for conversational agents, and which is used in the Google Assistant[6]. In our system (Fig. 1), we use DialogFlow not only to generate appropriate verbal responses, but also to assess what robot actions are most appropriate (see Sect. 3.2).

The essential components of a DialogFlow agent that users can define are Intents and Fulfillments. Intents define the relationship between the input (for example a spoke command to the robot) and the desired action. Creating Intents takes the form of defining a phrase (or set of phrases) that should be recognised, and then defining what should be done if a recognition occurs. Fulfillments allow the actions to be executed and are used to send commands to the robot (see Sect. 3.2).

One basic example of interactivity that can be defined in DialogFlow is smalltalk (Fig. 2). In the context of our system, smalltalk does not result directly in robot action, but can be extended to give a greater degree of verbal interactivity. This is one aspect that we focus on in particular in our engagement with participants.

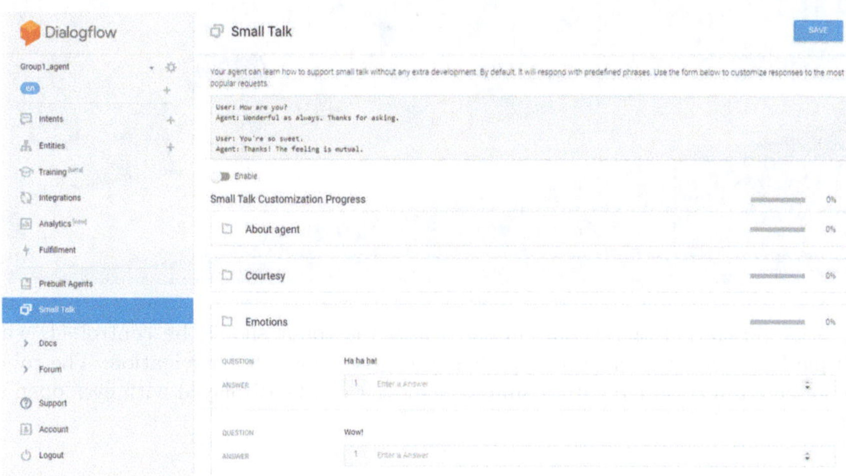

Fig. 2. A screenshot of the DialogFlow page showing the smalltalk definition area for one of the participant groups (Sect. 4.1).

While at present a freely available resource, the use of DialogFlow does pose a number of practical challenges, particularly if the system is used by children.

[5] https://dialogflow.com/.
[6] https://assistant.google.com/.

Primarily, a Google user account is required, which places restrictions on who can use the system, for example a minimum age restriction. There are however various circumstances in which children can make use of such services, particularly in educational contexts. This issue can also be circumvented by using an alternative dialogue planning system, or indeed a more basic hand-coded solution (in which all possible speech variations must be explicitly coded by hand), in place of DialogFlow. This is principle does not change the overall system schema, although it potentially reduces the inherent flexibility of the resulting system given that the background machine learning for interpretation of language input in the back-end of DialogFlow would no longer be used.

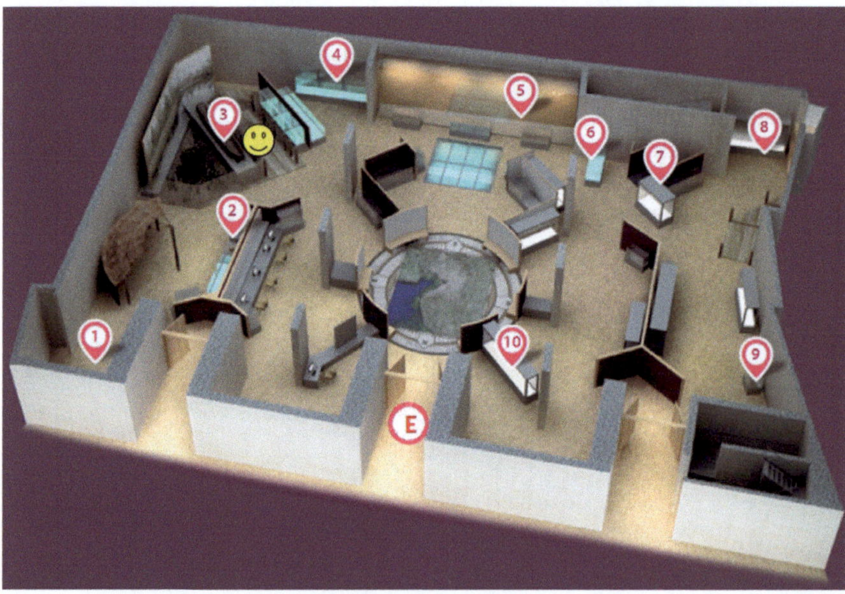

Fig. 3. A simulation of the museum in which the robot should be controlled, with multiple locations of interest shown that are targets for robot navigation. The robot is represented in this map with a smiley face (that can be displayed with eyes open or shut).

3.2 Fulfilment Server

The fulfillment server is an interface between the DialogFlow agent and the robot actions and events. A set of commands were predefined relating to robot movement around the museum space (e.g. go to one of 11 predefined locations, Fig. 3, comprised of 10 exhibits and the entrance), behaviours (e.g. open/close eyes), and the display of information (e.g. displaying information on the robot touchscreen, Fig. 4(a), or reporting the current location back to the DialogFlow agent). Importantly, due to the system setup (Fig. 1), these commands can be sent to either the real robot, or to the simulated robot (Fig. 3), with the difference simply being the definition of a webhook to access a web service hosted by the desired platform (Fig. 1). This allows the same DialogFlow agent to be easily repurposed.

3.3 Robot

The target robot platform for development in our application is based on the Scitos G5, extended with additional sensors and computing capacity. A mobile robot with low ground clearance, good stability, and standing at approximately 1.6 m tall, it is ideal for moving in indoor environments, and for safely interacting with people. It has a limited capacity for facial expressivity, extending to blinking eyes and manipulation of gaze direction (within a static head, see Fig. 4). The robot is controlled using the ROS middleware, which as a technical standard, facilitates integration with a range of specialist processing components (e.g. for navigation or visual processing) and interoperability with a range of other robot platforms.

Within the overall system however, the precise robot used is not important, provided that certain basic functionality is fulfilled, namely navigation and sufficient sensor capability for interaction with humans (e.g. microphone for speech recognition, and camera for person/face localisation). This arrangement, partially afforded by the use of ROS, means that a simulated version of the robot can be used in place of a real one (Fig. 3).

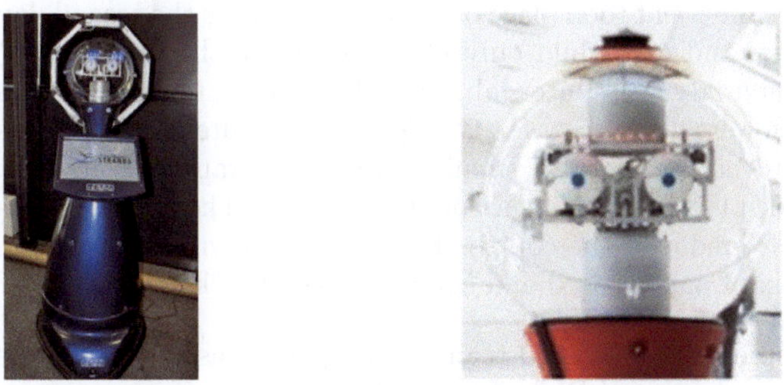

(a) A modified version of the Scitos G5 mobile robot.

(b) Some facial expressivity, namely eye blinking and gaze direction.

Fig. 4. The robot used for the case studies presented.

4 Case Studies

Using the system described above, we engaged learners in two events. Each of these were with different profiles of participants, in terms of both age and background, although the same system was used in both cases. In both cases, post-event questionnaires were given to the participants in order to gain insight into their experiences. Full details of these questionnaires and the responses are available in the appendix at the end of this paper.

4.1 School Children

Context: In a large scale STEM engagement event, we hosted nearly 200 10-year old children from primary schools in the local area over the course of two days (just under 100 on each day). The aim was twofold. Firstly to engage the children with technology and applications beyond their typical school experiences through engagement with robotics and artificial intelligence. Secondly to have the children involved in the design of interactive robot behaviours that could be potentially deployed on the real robot (the children were made aware of this possibility).

Participants: A total of 188 children from 14 local primary schools took part in this event. All participating children were in U.K. year 5, and were thus nine or ten years old, with the majority ten (due to the time of the school year this activity took place - mean 9.606, s.d. 0.505). The intention was that the activities would be targeted at schools whose pupils were of a lower socio-economic status, thus providing the children with an opportunity not normally available to them. There was an even balance of boys (50.7%) and girls (49.3%). The children were divided into a total of 36 groups, with group members within the same school, and resulting group sizes from 5 to 8 members.

Process: The event took place over two days, where the second day was a repeat of the first, and each child only attended one of the two days. The event was situated in a large computer lab, typically used by undergraduates. Each group of children were given access to two desktop computers, with the instruction that one was to be used for the simulation and programming, and the other to be used for displaying the instructions/worksheets. There were seven stages of the activity covered over the course of the day, each with an associated worksheet, with a mix of paper-based and computer-based tasks[7]:

(1) Group organisation and team name (paper-based, team name was applied to their developed DialogFlow system)
(2) Computer setup (intialising DialogFlow and the simulation)
(3) Assessing variability in speech (paper-based, thinking about people can say the same thing in different ways)
(4) Small talk (applying the speech variability ideas to small talk definition - see Fig. 2 for an example)
(5) Starting to define robot behaviours (Phrases and Fulfillments)
(6) Thinking about the strategic level of the desired robot behaviour

[7] The worksheets may be found at the following: https://sites.google.com/view/uol-ai-robotics/materials.

(7) Turning planned behaviours into Intents

Throughout the day, there was a robot present in the room to contextualise the eventual task (see Fig. 4) - various behaviours for this robot were demonstrated during the lunch break. At the end of the event day, the children were asked to complete an online questionnaire (using one of the computers available to their group), this was voluntary, and with the only identifiable information requested the name of the school they attended (for the purpose of further analysis by school - this is not covered in the present work). The children (and their teachers) were encouraged to continue development of their solutions from their schools: both the worksheets and the simulation setup remained online to facilitate this.

Outcomes: Qualitatively, both teachers and children verbally reported good levels of engagement and enjoyment throughout the event. This is supported by the post-event questionnaire, with a response rate of 76.6% (144 responses in total). Overall, the children reported enjoying the event (1.45 mean, 5-point scale where 1 is very positive), which is consistent with an overall positive opinion on robots (1.92 mean), and with a positive response to the idea of having a robot help them in various contexts, such as in a museum (1.63 mean), at school (1.69 mean), and at home (1.9 mean). The free-text responses provide some further insight, analysed in terms of keyword frequency analysis (please refer to the appendix for all the data). The term "Robot/s" were identified as the aspect most enjoyed, even though the presence of real robots was only a small part of their experience during the event. This is consistent with them identifying both "robot/s" and "programming" as aspects they learned, with these keywords appearing more frequently than when asked what they liked most about the day. In terms of negative aspects of the experience, where present (the keyword with the highest use was "nothing"), these mostly related to challenges of working in a group, rather than the robot activity per se.

Discussion: To a certain extent, it is relatively unsurprising that the children indicated that they had enjoyed the event, given that it involved a day away from school interacting with technologies they were previously unfamiliar with. It is interesting to note that the children identified "robot/s" and "programming" with greater frequency with respect to learning outcomes (122 and 61, respectively) than with respect to their enjoyment (81 and 43, respectively). Although this observation will require further investigation with respect to real learning/skill gains, it is an initial indication that the children were aware of being engaged in a learning process.

4.2 Undergraduate Students

Context: As part of a summer school (University of Valladolid, Spain), a two-hour activity was run for participants. The general topic was interactive robotics, with the aim of familiarising the participants with the associated methods and technologies. In this case, the focus was on expressive dialogue interactivity for (social) robots.

Participants: There were a total of 14 participants (and two professors), most of whom were undergraduate level. All were attendees at the summer school, and opted in to participate in the activity. According to a self-reported level of experience in the topic, there was only middling pre-existing familiarity declared (3.37 mean on a 5 point scale, 1 being high familiarity), consistent with the undergraduate level.

Process: The activity was essentially a compressed version of the day-long activity presented to children (Sect. 4.1), though with the emphasis on individual work, rather than group work. Due to the location of the workshop, there was no physical robot present, and so only the simulation environment was used. However, the same principles applied as for the children's event.

Outcomes: As with the children's event, self report feedback forms were administered to all participants (with full results shown in the appendix): eight participants returned these. A high level of satisfaction with the event (1.63 mean, 5-point scale where 1 is positive) was reported, and an agreement with both the format of the activity (1.63 mean) and the content of the activity (1.75 mean). Qualitatively, a good level of engagement with the delivery team was experienced, consistent with the feedback from responses.

Discussion: While involving participants at a higher level of education and experience, this case study demonstrates the applicability of our dialogue design for mobile robots system beyond children. The emphasis of such an engagement naturally changes, focusing more on implementation details than the foundational concepts, but the system itself remained identical to the previous case study, as indeed did the principles of the problem presented to them with the focus on dialogue interactivity for real robots.

5 General Discussion

In this paper, we have motivated and presented a system to engage learners with the design of interactive systems, in our case specifically to increase the social interactivity of museum guide robots. This system employs a freely available commercial interface for designing behaviours, but has been interfaced to a back-end that facilitates the execution of behaviours seamlessly on both real and virtual robots. Two case studies were described: one large scale involving nearly 200 ten year old children, and one smaller in scale involving 14 undergraduate students. Despite the large disparity in participant groups, in both cases the same system was used (Sect. 3). The metrics we employed to inspect engagement and enjoyment (self-report measures) demonstrated success for both. We are clear that these metrics can only be interpreted as *indicators* for engagement with the learning content of the respective activities - however, the data obtained is consistent with this perspective.

What we have not yet done as part of this research is make any assessment of learning outcomes, beyond the immediate self-reflection of the children. Since we were concerned with skill development and engagement, this would have necessitated both a longer-term perspective, and a clearer notion of definitive outcome for the respective participants. Since this work is ongoing, with a final engagement event for the first case study with children yet to take place (approximately six months after the initial activity days), one aspect that we intend to inspect is the relative complexity of the solutions (i.e. DialogFlow agent behaviours) developed by the teams of children. While this would not be a complete means of assessing learning outcomes, it would provide insight into continued engagement beyond the initial activity, and relative mastery of the tools available to them.

Our system and application to the two events have demonstrated consistency with the principles of previous applications of robotics to education. What we further emphasise here, through the tie-in to a real application domain, is the link to ongoing research efforts. This has the intention of being mutually beneficial, and adding a level of motivation/engagement for the learners as a result.

There are aspects of the system that could be extended to take further advantage of the features we identify as central to the approach proposed in this paper. Primary among these would be the provision of a higher-fidelity simulation environment that is similarly accessible through a web browser. This would facilitate both a greater immersive experience when using the simulator, but also a more realistic view of the defined robot behaviours, which in turn would allow a greater degree of detail in the designed interactive behaviours. Ensuring accessibility through a web browser, as we have currently achieved, would expand the set of the general public (including schools) able to engage with these activities, as no specialist software installations are required. If implemented in such a way as to maintain compatibility with a real robot platform (which our use of ROS ensures is a straightforward task), then we would maintain the ability to view the designed behaviours operate in the 'real world', thus adding the application dimension to the learners experience - just as we intend to achieve for the children designing dialogue interactivity for the museum guide robot.

6 Conclusion

Our intention in this paper was to describe a system we have developed for engaging learners in a wide range of contexts with the design of dialogue-based interactivity for mobile robots, and more specifically in this case, with application to the heritage domain. With an emphasis on a web-based solution that is grounded in both a real robot system and a real application domain, our intent is to enhance the benefits to both driving research through potential user-group engagement, and enhancing motivation by providing a real application context for the learners involved. In a practical sense, the system is scalable to large numbers of simultaneous learners (as shown in the first case study for example), but also to a wide range of learners (as shown by the very different demographics of the participants of the two case studies). Our observations from preliminary events support the view that the system is both usable and successful in supporting engagement with the problem presented to the participants, with indications that this can persist over an extended period of time. It remains to be seen whether such high engagement will result in tangible learning outcomes/benefits, however, we have put in place a system that facilitates such efforts.

Appendix

Child Learner Questionnaire Data

Questions asked of the children in the questionnaire (see Sect. 4.1) are as follows, with the mean responses (and *s.d.*) where appropriate (144 responses unless otherwise stated):

(1) basic demographics (age, gender, school)
(2) have you ever used real robots (no, yes at home, yes at school, yes somewhere else)
(3) how do you feel about robots in general (5-point scale, 1: very positive, 5: very negative): 1.92 (*0.98*)
(4) how would you feel about having a robot in the museum to help you? (5-point scale, 1: very positive, 5: very negative): 1.63 (*0.92*)
(5) how would you feel about having a robot at school to help you? (5-point scale, 1: very positive, 5: very negative): 1.69 (*1.2*)
(6) how would you feel about having a robot at home to help you? (5-point scale, 1: very positive, 5: very negative): 1.90 (*1.32*)
(7) have you enjoyed the day today? (5-point scale, 1: yes very much, 5: not at all): 1.45 (*0.84*)
(8) have you ever been to the Collection Museum (in Lincoln) before? (yes/no): 27.8% yes
Keyword analysis was applied to responses of the free-text questions (below): a simple frequency count was employed, excluding common words (*a, to, the, and, lot, that, but, is, are, of, it*).
(9) What do you think you have learned today? 137 responses, 7.45 mean words per response (*s.d.: 4.45*). Top keywords (with >10 uses): robot/robots (122), how (78), program/programming (61), I (23), about (22), learnt/learned (18), do (16), have (12), you (10).
(10) What did you like the most about today? 138 responses, 5.43 mean words per response (*s.d.: 3.77*). Top keywords (with >10 uses): robot/robots (81), program/programming (43), seeing (16).
(11) What did you not like about today? 132 responses, 5.31 mean words per response (*s.d.: 4.69*). Top keywords (with >10 uses): nothing (65), I (23), all (13), we (11).

Workshop Feedback Data

A feedback questionnaire was provided to the participants after completion of the workshop (see Sect. 4.2), with the questions and mean responses (and *s.d.*) shown below. All questions were Likert items with a five point scale (1: positive, 5: negative). Eight questionnaires were completed.

(1) Indicate your previous knowledge about the topics covered: 3.57 (*1.27*)
(2) Are the objectives of the activity clear and concrete? 1.75 (*1.16*)
(3) Have the contents been useful? 1.75 (*1.16*)
(4) Have the resources and documents provided been adequate? 1.88 (*1.46*)
(5) Have the teacher's explanations been clear? 2 (*1.31*)
(6) Does the teacher dominate the topic of the activity? 1.63 (*1.41*)
(7) Has the format of the activity been adequate for the topic addressed? 1.63 (*1.06*)
(8) Your overall satisfaction with the teacher is: 1.63 (*1.41*)
(9) Your degree of overall satisfaction with the activity performed is: 1.63 (*1.06*)
(10) Would you recommend carrying out this activity? (7 responses): 1.29 (*0.49*)

References

1. Baxter, P., Ashurst, E., Kennedy, J., Senft, E., Lemaignan, S., Belpaeme, T.: The wider supportive role of social robots in the classroom for teachers. In: 1st International Workshop on Educational Robotics at the International Conference on Social Robotics, Paris, France (2015)
2. Cakmak, M., Takayama, L.: Teaching people how to teach robots: the effect of instructional materials and dialog design. In: HRI, pp. 431–438. ACM/IEEE, Bielefeld (2014)
3. Chowdhury, G.: Natural language processing. Ann. Rev. Inf. Sci. Technol. **37**, 51–89 (2003). https://doi.org/10.1002/aris.1440370103
4. Del Duchetto, F., Kucukyilmaz, A., Iocchi, L., Hanheide, M.: Do not make the same mistakes again and again: learning local recovery policies for navigation from human demonstrations. IEEE Robot. Autom. Lett. **3**(4), 4084–4091 (2018). https://doi.org/10.1109/LRA.2018.2861080
5. Diprose, J., MacDonald, B., Hosking, J., Plimmer, B.: Designing an API at an appropriate abstraction level for programming social robot applications. J. Vis. Lang. Comput. (2016). https://doi.org/10.1016/j.jvlc.2016.07.005
6. Fischer, K., Lohse, M.: Shaping naive users' models of robots' situation awareness. In: RO-MAN 2007 - The 16th IEEE International Symposium on Robot and Human Interactive Communication, pp. 534–539 (2007). https://doi.org/10.1109/ROMAN.2007.4415144
7. Hanheide, M., Göbelbecker, M., Horn, G.S., Pronobis, A., Sjöö, K., Aydemir, A., Jensfelt, P., Gretton, C., Dearden, R., Janicek, M., Zender, H., Kruijff, G.J., Hawes, N., Wyatt, J.L.: Robot task planning and explanation in open and uncertain worlds. Artif. Intell. **247**, 119–150 (2017). https://doi.org/10.1016/j.artint.2015.08.008

8. Horn, M.S., Solovey, E.T., Crouser, R.J., Jacob, R.J.: Comparing the use of tangible and graphical programming languages for informal science education. In: Proceedings of the SIGCHI Conference on Human Factors in Computing Systems, CHI 2009, pp. 975–984. ACM, New York (2009). https://doi.org/10.1145/1518701.1518851
9. Karreman, D.E., van Dijk, E.M.A.G., Evers, V.: Contextual analysis of human non-verbal guide behaviors to inform the development of FROG, the fun robotic outdoor guide. In: Salah, A.A., Ruiz-del Solar, J., Meriçli, Ç., Oudeyer, P.Y. (eds.) Human Behavior Understanding, pp. 113–124. Springer, Berlin (2012)
10. Kopp, S., Gesellensetter, L., Krämer, N.C., Wachsmuth, I.: A conversational agent as museum guide–design and evaluation of a real-world application. In: Intelligent Virtual Agents, pp. 329–343. Springer (2005)
11. Lee, C., Jung, S., Kim, K., Lee, D., Lee, G.G.G.: Recent approaches to dialog management for spoken dialog systems. J. Comput. Sci. Eng. **4**(1), 1–22 (2010). https://doi.org/10.5626/JCSE.2010.4.1.001
12. McNerney, T.S.: From turtles to tangible programming bricks: explorations in physical language design. Pers. Ubiquit. Comput. **8**(5), 326–337 (2004). https://doi.org/10.1007/s00779-004-0295-6
13. Papert, S.: Mindstorms: Children, Computers, and Powerful Ideas. Basic Books Inc., New York (1980)
14. Šabanović, S.: Robots in society, society in robots. Int. J. Soc. Robot. **2**(4), 439–450 (2010). https://doi.org/10.1007/s12369-010-0066-7
15. Thrun, S., Bennewitz, M., Burgard, W., Cremers, A.B., Dellaert, F., Fox, D., Hahnel, D., Rosenberg, C., Roy, N., Schulte, J., Schultz, D.: MINERVA: a second-generation museum tour-guide robot. In: Proceedings of the 1999 IEEE International Conference on Robotics and Automation IEEE (1999)

Educational Robots with Arduino: Annotated Prototypes

David Cuartielles[1(✉)], Nerea Iriepa[2], Carlos Rodriguez[2], Ernesto Lopez[2], and Jose Garcia[3]

[1] Malmo University, 20506 Malmo, Sweden
david.cuartielles@mau.se
[2] Arduino AB, 21119 Malmo, Sweden
{n.iriepa,c.rodriguez,e.lopez}@arduino.cc
[3] Abierto Project, 21119 Malmo, Sweden
j.garcia@abierto.cc

Abstract. This paper introduces the experiments realized by Arduino Education in the field of educational robotics. The paper, written as a collection of annotated exemplars, covers a series of prototypes, kits, and full educational programmes which were tested with students of different ages and educators. Some projects are of a do-it-yourself (DIY) nature, a property we came to describe as DIY-ness, while some others have been manufactured and served to tens of thousands of students. There are however things in common that can help others in the conceptualization, development, and deployment of educational robotics initiatives.

Keywords: Digital manipulatives · Robotics kit · Education programme

1 Introduction

Educational robotics is, without dispute, one of the strongest trends in contemporary education across the world. Whether emerging at local level through committed educators and early adopting schools, or imposed through governmental initiatives, the use of mechatronics in formal education at primary, secondary and high school education is becoming more and more common. In parallel, technology is evolving rapidly, offering new opportunities in the field of education, but also challenging the way we understand society and all sorts of activities in everyday life.

This paper describes several experiences realized over the course of eight years offering interaction designers, engineers, educators, and policy makers insights to how different educational experiences took form. These experiences were designed to accommodate different contexts and taking into account socioeconomic aspects as a main driver.

In a sense, this paper could be considered an annotated portfolio - a concept that will be explained later - in the field of educational robotics, a catalogue for others to look into and learn from our mistakes, but also our successes. One of the main aspects to observe is that all designs were the result of co-design processes, a design technique

M. Moro et al. (Eds.): Edurobotics 2018, AISC 946, pp. 161–174, 2020.
https://doi.org/10.1007/978-3-030-18141-3_13

that, as will be discussed later, offers both advantages while also constrains the generalizability of the design.

1.1 Limitations

This work presents different devices co-designed between students and teachers at Malmo University, but also between Arduino AB – a company dedicated to the creation of educational programmes – and members of its community. Some of the robots and programmes were redesigned for production purposes, manufactured, an even sold by Arduino AB, in Sweden. It should be mentioned that it is not the intention with this paper to advertise products, but to explain the circumstances around the different designs, what worked in the interaction with users, and which were the tests made with the robots in real life scenarios, if any. It should also be noted that most of these objects are not for sale any longer, or they never were, as they were mere prototypes or even Inclusive/Multiple Prototypes tested with hundreds to thousands of users [1].

2 Theoretical Framework/Methodological Attitude

Annotated portfolios are collections of experiences put together and including enough explanations to them so that it is possible to draw a "red thread" between them. It is a technique suggested in Research through Design as described by Gaver [2].

The approach to the collaboration between Malmo University and Arduino AB in the field of educational robotics over the course of several years has its roots in the constructionist education paradigm [3, 4] and its hands-on approach [5] to learning that is close to the reflection in action theories [6]. Reflection in action establishes that there is a learning process that happens during the interaction in the field. This approach is two-fold when it comes to educational robotics: on the one hand we believe that it is not possible to create educational experiences of any kind without interacting in the field and performing tests with users. On the other hand, due to the physical nature of robots, the tests need to include prototypes and software for people to experiment with.

3 Experiences

This section is dedicated to the description of experiences that have the common denominator of having been co-created by designers, engineers and educators, that have been orbiting around Malmo University (MaU) and Arduino AB (ARD) over the last ten years. Both have collaborated in formal and informal settings. Some students from MaU do interships at ARD, while ARD sometimes collaborates with MaU in EU research projects, courses, exchanging materials, and even bringing in state of the art prototypes into the classroom for students to evaluate.

Each one of the following subsections will introduce the context of the design, the design intentions, and a summary of the process of building the design. The final conclusions will help to compare the different projects and discuss about certain aspects like the sustainability of each one of the objects.

3.1 OhOh – Born to Be Hacked

During a 4 years long residency at the cultural centre *Fabrica de Artes y Oficios FARO de Oriente* in Mexico City, we worked in re-launching a Computer Clubhouse space that had not been operative for an undetermined amount of time. Our role was to get the existing non-functional computer equipment to work, as well as to create educational content and programmes using open source hardware and free software components. After a 2 years process, once the basic aspects of the space had been put into motion (working computers, Internet, sound studio, projection facilities, courseware) and once we found a person to carry on the work on a day to day basis, we opened a process of co-creation of new activities with the local users, who were 6 to 18 years old children, and teenagers. This implied looking at what had been done until then in terms of content, what was available in the Clubhouse's material library, and what opinions the kids had about what they wanted to explore next (Fig. 1).

Fig. 1. OhOh robot collection after being modded by kids at FARO de Oriente, Mexico, 2010.

We arranged a simple survey to ask what was that the kids wanted to do and the resulting answer was "robots." This almost unanimous response put us under the situation of having to create an educational robot that could be both affordable in terms of price and materials. These design constrains forced the creation of a strategy based on the concepts of: locality, cost, and ease of implementation. In other words, in order to make the robot as cheap as possible, we had to make sure that the design was easy enough for the Clubhouse's responsible to manufacture by himself. This could allow him having some of the older participants to help out. It would also allow for any kid to implement her personal robot in case she wanted [7].

The robot was made sourcing parts at local stores, what required a major research operation, having to look for places where to obtain wheels, motors, electronic components, and the copper boards and chemicals needed to etch circuit boards. In parallel, back at our lab, we studied how to create a series of exercises for the kids to learn the basics about robotics, built software libraries to simplify the operations with the robot, etc.

We introduced the robot to the Clubhouse by arranging a course where both the leader and the kids would learn about how to use the robot and how to run educational activities with it. The first time we manufactured 15 robots using pre-etched PCBs, to later explain the Clubhouse educator how to manufacture his own PCBs from scratch. We ran two different sets of workshops together with the Clubhouse as a way to make some knowledge transfer and collect feedback about the designs. The combination of price and features was very attractive, what together with the DIY-ness feature of the robot, made it suitable for a mid-term educational process. The lack of an out-of-the-box experience made this process hard to replicate for novices without the supervision of a mentor.

3.2 Designing the Arduino Robot

The Arduino Robot [8] was an 18 cm diameter, 18 cm high, two-wheeled, circular robotic platform inspired by the robots used at Robocup Junior's robotic soccer competitions. It was created as a collaboration between the Complubot robotics association – from Alcala de Henares, Spain – and Arduino. Our goal was to provide the robotics education community with the so-called Arduino out-of-the-box experience: for a novice in the field it should take a minimal amount of time to get started with performing all sort of experiments.

The design included two different processors: one to control the motors and read line following sensors, and yet another one to handle the user experience. This second processor was the one that would be directly accessible for users to program. The first one would act as a peripheral to the latter and came pre-programmed from the factory. The robot included a series of features that were expressed in the form of libraries of code. On the robot, there was a magnetometer, an LCD screen, connectors for infrared or ultrasound distance sensors, connectors for servo motors, a battery recharging circuit with voltage measuring features, a line following sensor made of an array of 5 infrared sensors, a sound-speaker, a keypad, a potentiometer, an infrared remote-control receiver, and a dual half-bridge controller with current sensing (Fig. 2).

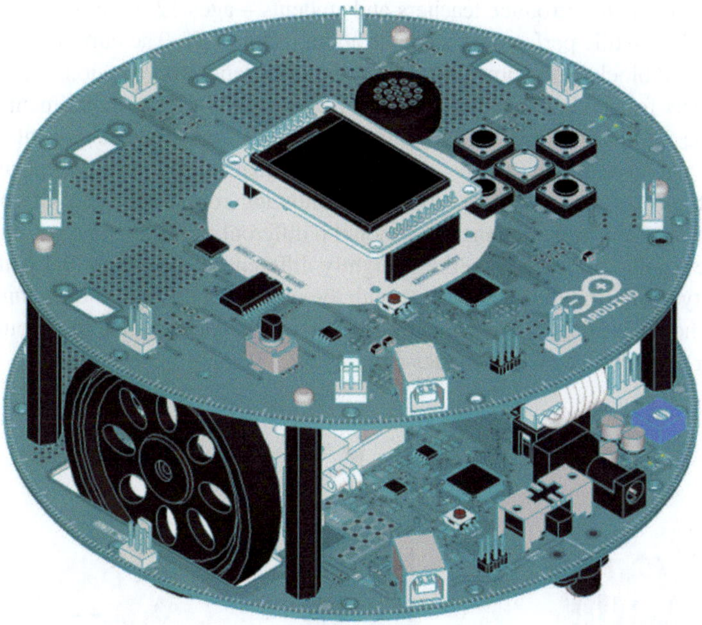

Fig. 2. Illustration art from the Arduino Robot, courtesy of Arduino AB, 2013.

All of the above-mentioned features were accessible to users as object-oriented code for them to create their own behaviors with the robot. It, however presented a challenge during the robot configuration phase directly after unboxing. The difficulty arouse from the high variability of the motors used in the design. We made many compromises trying to keep the robot's price as low as possible. For example, we chose DC motors without encoders what made hard to align the robot's wheels. We compensated this with a complex calibration procedure, as well as a proportional–integral–derivative (PID) controller implemented at library level using the magnetometer and the current sensing coming from the motors. While this approach is perfectly valid, it forced a certain way of programming the robot that was not as flexible as expected by the average educator. This made the software production for the robot complicated.

While the design caused a good impression in the Arduino educational community (a couple of thousands of robots were sold), it was not enough to keep it in the portfolio of Arduino's educational products as there was not a right balance between price, manufacturing complexity, and software limitations.

3.3 CTC Robots That Walk, Robots You Tickle

Creative Technologies in the Classroom (CTC) is an educational programme [9], a full curriculum for a class of 30 students including all the materials in the form of a kit needed to perform 26 different mini-projects, lasting between 6 and 9 months (depending on the amount of sessions devoted per week), and offering webinars and online support to the teachers and schools participating in the experience. CTC's main pedagogic goal is to introduce teachers and students – ages 12 to 16 – to project-based learning (PBL) while performing a course about technology. The curriculum is divided in educational blocks and comes with a set of guided laboratory sessions and suggested mini-projects for having 6 groups in class performing different experiments for each block. One of those blocks is dedicated to robotics and students are introduced to somehow unconventional robots: a clown robot that walks, a rocking-chair-sitting-robot that shivers when tickling it, a light following robot, etc. The kit comes with enough components to simultaneously mount 6 different projects – one per group. Each one of the teams will learn something slightly different to the others and will have the opportunity of sharing their experience with the others. At the end of each mini-project, teachers should arrange a presentation session where each one of the groups should pitch their project to their colleagues (Fig. 3).

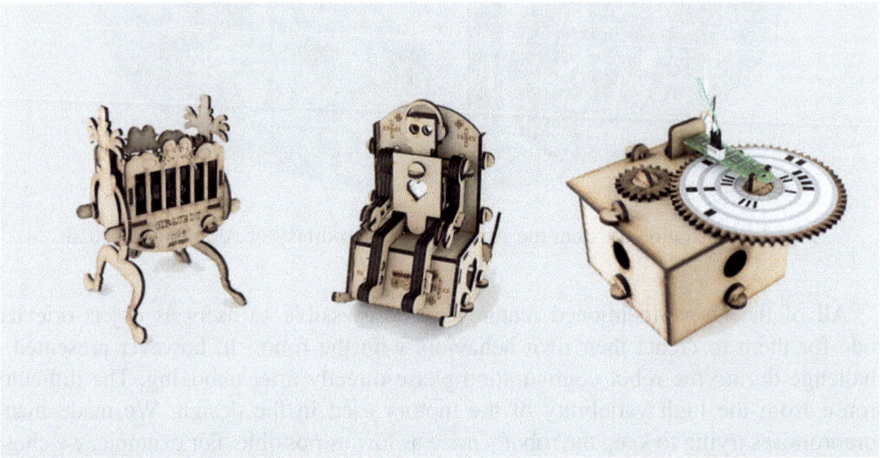

Fig. 3. Walking robot, tickle robot, and binary LP, courtesy of Arduino AB, 2016.

We looked at how to make this kit of 99% standard off-the-shelf electronic components, however we designed our own infrared line following sensor, just to simplify the process of mounting sensors on a breadboard, thus accelerating the class' pace. The kit includes mechanical parts made in HDF synthetic wood, what makes the pieces durable, but at the same time replicable and augmentable. Yet another key aspect is the contact with the teachers through a specially made platform. The content is licensed through a Creative Commons license – what allows the schools to make copies and modifications of all of it – and it is made available through an online platform that gives them access to the latest improvements, webinars with experts, access to a teacher-only forum, and direct calls with a dedicated Arduino support specialist. This project has been tested over the course of 5 years (2013–2018) and reached over 17.000 students by the hand of 800 teachers. Therefore, the key aspects of this design are: modularity understood as reusability of parts, standardization of components, and direct access to support for educators.

3.4 Svante, Reaching Out

While the DIY-ness of the OhOh robot is attractive for some, it is a too-steep entry level for many that keeps educators away from experimenting with robots in the classroom. At the same time, high prices combined with technical challenges – as happened with the Arduino Robot – makes the design non-desirable. Scalability is a very important feature in education: classes vary in size, and it is important to be able of shrinking or enlarging the class; this is something we experimented with during the CTC project.

During 2014, Arduino's educational team went into making an exercise of self-critique and tried to imagine ways to gather the best aspects of the OhOh, Arduino Robot, and CTC projects. We wanted to prototype a new design that would allow us making mid-size experiments in a class. Thus, we created the Svante robot. Technically, Svante was a minimization of the Arduino Robot, included the line following sensors, connectors to augment the board with other components, but just one processor, and with a simplified set of libraries (only 2). This design would allow answering the question of how a class would work if each and every one of the students had her own robot (Fig. 4).

Fig. 4. View of the Svante robot modded by kids, courtesy Arduino AB, 2014.

One robot per kid was the main driving force behind the creation of the Svante robot, a prototype created and deployed in the hundreds during a summer course experience in the city of Zaragoza, Spain. Svante was a hackable robotics platform that was pre-made, aimed at running 1-week-long workshops with students where they would devote 4 h per day to learn about electronics, mechanics and software. The design included a water-drop-shaped two-wheeled robot, with enough extra off-the-shelf components to remote-control it via infrared, add mechanical extensions to it, or extra lights and sound.

We designed a cardboard platform, to be placed on top of the electronics, and a series of mechanical parts (also in cardboard) for the students to experiment in making their own designs, decorate the robot, and not mind about breaking something, as it was easy to replace. This added a component of open experimentation directly with the robot itself. The design of the courseware for the experiment was key as well as its deployment. In total we deployed 150 robots that were given away in three groups of 50 to kids participating in the experience for a week. This means that we had to design educational material for the educators that would collaborate with us. The 50 weekly participants were divided in groups what brought us to educate a total of 10 mentors to be able of supporting the activities with the robots. There were two mentors per group of kids, and there were different activities taking place, not just robotics. The mentors had to rotate and eventually be at other activities, therefore all mentors had to get trained even if they might not specifically be part of the robotics teaching.

While there are some similarities in terms of the teacher curriculum preparation between CTC and the Svante experience, we didn't offer the same level of online support, nor webinars, since the activities didn't last long enough to benefit from that kind of action. We instead focused in making a two days training where the mentors were introduced to all of the activities at an accelerated pace.

This activity created a group of young roboticists that decided to continue to meet over one year to see what they had been doing with their robots. The culture center where this activity took place hosted meetings where the kids came – hand by hand with their parents – to share their findings and make new experiments and competitions. Seems like owning the educational device contributes to the follow up experimentation home, which is also part of the personal education space.

3.5 LCR a Design Experiment

LCR is the acronym of Low Cost Robot, an experiment at Arduino dedicated to the creation of a robot in the form of a shield to the most extended Arduino board: the Arduino UNO. Only a couple of prototype robots were manufactured. We entertained the idea of having a system to allow schools, that already had been experimenting with electronics, augmenting their existing class materials to run simple robotics experiments. The experience of the Svante project proved that having a maximal amount of devices per class is a desirable scenario, as it encourages the students to continue experimenting back home. At the same time, the price of the materials should be as affordable as possible so that educators could adjust their inventory to the class' size (Fig. 5).

Fig. 5. LCR robot prototype, courtesy of Arduino AB, 2017.

LCR was created as a shield that also revised the technical flaws observed during the Svante experiment, namely the choice of motors and – yet again – the lack of encoders to properly determine the movement of the wheels in comparison to the others. Therefore, it included access to motors with encoders, 4 neo-pixel RGB addressable LEDs that could be used for indications, infrared sensors, and connectors for external sensors. The robot shape was designed for it to move in 3 different ways: normal, upside-down and vertical. By using one of the latest Arduino boards (101 or WiFi Rev2) it should be possible to perform experiments using the embedded communication chips and MEMS sensor, in order to e.g. design a simple inverted pendulum robot. Price was, once more, a strong diver for this design.

The one thing that was made differently in this project was that we didn't rely only in the use of the Arduino IDE to program the firmware for the robot. We created our own visual programming language (VPL) extension to the IDE that generated C code by dragging and dropping simple visual blocks on a user interface. The reason to create this VPL as a tool for the Arduino IDE is to simplify its usage as well as maintenance for future revisions. The IDE has a version control mechanism that automatically detects updates and suggests them to the users upon booting the software.

3.6 Escornabot, the Community's Response

The experiments described up to this point focused in the creation of robots to be used in the second part of primary as well as secondary education cycles. It is however a challenge to reach younger audiences. A powerful alternative comes from the Arduino community itself. While it is possible to find hundreds of experiments in robotics, it is hard to find initiatives that go beyond the mere creating of kits as a way to have products people could buy. There is where the Arduino community made a difference by creating, supporting, and maintaining, a family of open source and inexpensive educational robots called Escornabot [10].

This project was born from the need of having a robot like the commercial Bee-Bot® targeting young audiences and introducing them to robotics through the exploration of spaces via maps and other pedagogical constructs. It was key to this project to identify motors that were easy enough to drive (5 V stepper motors in this case) to realize the cheapest – both in terms of price and computation – robot possible. The design combines DIY qualities as it is meant to be a kit that teachers could purchase and assemble by themselves. The mechanics to the robot are 3D printed and the intelligence can be programmed into an Arduino Nano or Micro (there are versions of the robot that actually support both types of boards). The key aspect of this robot is price. It is possible to acquire a fully assembled robot for about 50 Euros, what makes it very affordable for any circumstances. It is being broadly used by early adopters that are evangelizing its use and publicly sharing educational resources around the design. Because it is open source, there are at least 5 different sources where to acquire parts to make one of these robots (Fig. 6).

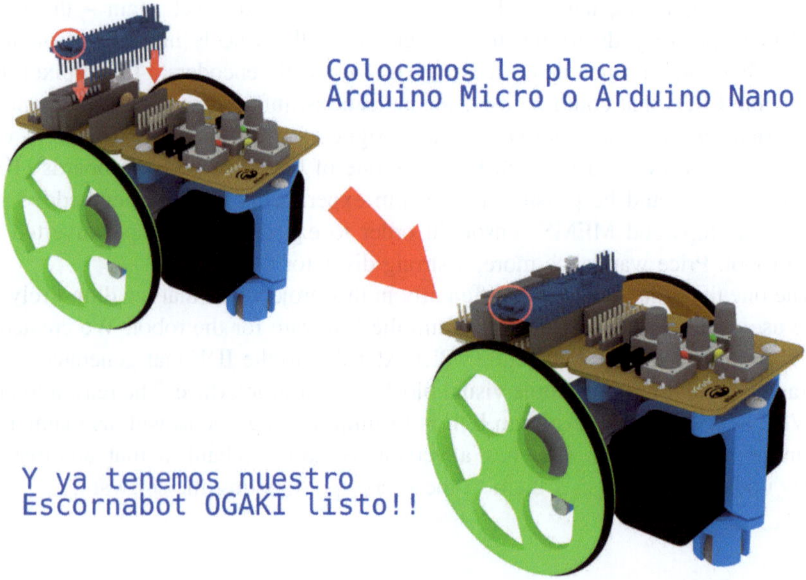

Fig. 6. Excerpt, construction manual, Escornabot Ogaki, courtesy of the Abierto Project, 2018.

In a way, this robot is being successful and opening up for other experiences to happen on top of it. At this point it represents a design opportunity as there are many improvements that could be made, and the community is bringing some of these improvements on a weekly basis. Currently there are educators and makers in Europe (mainly Spain) and LatAm (mainly Mexico) creating revisions of the designs and content and sharing everything online.

3.7 A Kit for Engineers

In the period 2016–2018 at Arduino there was work being done in understanding which were relevant experiments in technology done at engineering universities in the field of mechatronics. It was the goal to produce a new type of kit – the Arduino Engineering Kit – that would allow students experiment with state of the art software tools like MATLAB® which is broadly used both in educational and professional contexts [11]. We took into account the experience of designing different types of modular systems, including all of the projects mentioned above, and looked at the creation of a modular experimental set that could be used by students individually or in groups to learn basic aspects of motor control (PID), computer vision, creation of state machines, and how to translate mathematical equations from the world of physics to the world of software (Fig. 7).

Fig. 7. Building the balancing motorcycle experiment, courtesy of Arduino AB, 2018.

The experiments use the Arduino MKR1000, a design that includes WiFi communication, what allows the board talking directly to MATLAB. This helps making all sort of experiments where there is a direct interaction between the software and the physical implementation of the experiment. Also, it is possible to write firmware for the Arduino board using the VPL inside MATLAB (called Simulink®) what offers a very interesting alternative to programming Arduino boards. The kit presents three different complex projects. Each one of them makes use of a *carrier board* that can control up to 8 motors simultaneously. This carrier board has the same processor as the MKR1000 and comes with an open source specialized firmware aimed at controlling the motors automatically as it has its own PID algorithm implemented.

The experience required looking at what is being told at courses in engineering all over the world and find out a way to generalize some of that knowledge in the form of a reusable kit, that includes enough parts to realize the three projects: a computer vision-controlled rover, a drawing robot hanging from a wall, and a two wheeled self-balancing motorcycle. The kit itself is supported by educational material available online. However, the contents have been designed in an open-ended way, having the goal of inviting the students to go beyond the offered content, thus fostering their curiosity. It is still soon to write about the outreach of this project, as it was launched barely in May 2018. It is worth mentioning it here as it is our latest iteration in the world of designing educational robotics components and experiments, and to set up a full picture of the work performed over the last 8 years.

4 Common Ground – Lessons Learned

The above-mentioned experiments, programmes, and kits have a clear educational intention. However, they are different from one another as some of them try to address students directly (like Svante, LCR, or the Engineering Kit), while others are more focused in helping the teachers introduce complex technological concepts into a certain demographic (OhOh, Escornabot, or CTC). The cases were created with modularity and openness in mind, it is possible to contribute to the projects by adding content, software, and even mechanical enhancements to any of them. Some of the projects have been relatively successful, while some others never left the prototype stage at the laboratory.

Some key findings and learned lessons that can be extracted from the different experiences presented in this paper are:

- There is not a single way to approach educational robotics in a scaled-up fashion. Different strategies are needed based on age, teacher experience, and socio-economic aspects.
- Teacher support is not always the answer, sometimes excellent documentation will be better. Longer processes work better with group and personalized teacher support.

- The amount of materials matters, group experiences require less materials, but better planning and coordination between groups.
- Future classes will have different speeds, not all the kids will have the same knowledge in the subject by the time they start the class what will condition the kind of educational programmes we design.
- Content has to be complex enough to represent a challenge to the course participants. Complexity has to be leveled to the circumstances and unpacked by the use of educational paths. The design of those educational paths is crucial and should be iterated, e.g. the CTC programme is under constant revision including feedback from participating teachers.
- We tend to design for student engagement, but teacher engagement is equally important. Teachers are the main vehicle to education, we need to think about classroom management and sometimes it is better to think in terms of programmes instead of kits.
- If possible, activities should continue outside the classroom. If all students can have their own materials and bring them home, they will get the chance to rehearse and learn at their own pace, something that is much harder to achieve in a laboratory.

There is a common aspect to the projects in the paper and it is that it is possible to quickly create educational robotics platforms thanks to the existence of open source hardware and software components by either intervening the existing hardware or by using off-the-shelf components. In a matter of months, one could create a whole educational curriculum using standard parts, but in order to really have something to scale, at the time of designing we need to think about modularity in terms of technology, standardization in terms of interaction with users, and generalization of content. Future educational platforms need to sort out how to deploy experiences in a way that they could support as many people as possible. Making the technology is not enough, contemporary education needs technological maintenance as well as support to educators. In a sense, the maker world needs to professionalize its offering if it wants to success in the space of education.

References

1. Cuartielles, D.: Platform Design: Creating Meaningful Toolsets Where People Meet. Malmö University Press, Malmö (2018)
2. Gaver, W.: What should we expect from research through design? In: CHI 2012 Proceedings of the SIGCHI Conference on Human Factors in Computing Systems, pp. 937–946 (2012)
3. Papert, S., Harel, I.: Situating constructionism. In: Constructionism, pp. 1–11 (1991)
4. Falbel, A.: Constructionism Tools to build (and think) with (1993)
5. Resnick, M., Martin, F., Berg, R., et al.: Digital manipulatives: new toys to think with. Commun ACM 281–287 (1998). https://doi.org/10.1145/274644.274684
6. Schön, D.: From technical rationality to reflect ion-in-action. In: The Reflective Practitioner: How Professionals Think in Action, pp. 21–69 (1983)
7. Cuartielles, D., Yang, X.: Oh_Oh Experiments With Robots – Medea (2010). http://medea.mah.se/2010/08/oh_oh-experiments-with-robots/. Accessed 25 Feb 2017

8. Cuartielles, D., de la Riva Iriepa, N., Gallego, I., et al.: Arduino Robot (2011). https://www.arduino.cc/en/Main/Robot. Accessed 4 Jan 2017
9. Arduino CTC 101. https://create.arduino.cc/ctc/101/. Accessed 31 Jul 2018
10. Escornabot. http://escornabot.com/web/. Accessed 31 Jul 2018
11. Arduino Blog: Arduino goes to college with the new Arduino Engineering Kit! https://blog.arduino.cc/2018/05/12/arduino-goes-to-college-with-the-new-arduino-engineering-kit/. Accessed 31 Jul 2018

Flexible Techniques for Fast Developing and Remotely Controlling DIY Robots, with AI Flavor

Dimitrios Loukatos[1,2(✉)], Ken Kahn[3], and Dimitris Alimisis[2]

[1] Department of Natural Resources Management and Agricultural Engineering, Agricultural University of Athens, Athens, Greece
dloukat@gmail.com

[2] EDUMOTIVA - European Lab for Educational Technology, Athens, Greece
alimisis@edumotiva.eu

[3] Department of Education, University of Oxford, Oxford, UK
kenneth.kahn@it.ox.ac.uk

Abstract. During the last years we are witnessing a very successful osmosis between innovative and cost-effective credit card - sized computers and education. These computers, equipped with low cost sensors or actuators, can be the "heart" of various DIY robotic artefacts. This environment allows for a mixture of thinking and making activities that can be very meaningful in terms of pedagogy and science. Indeed, similar practices, usually referred as STEM or STEAM activities, are applied in many educational institutions, from primary schools up to universities, with most of the effort to focus on secondary school students. The overall process, although promising at the beginning, is not always straightforward to keep up with. More specifically, as students get more experience, they develop a hunger for more complicated scenarios that usually demand features like remote interaction with simple Artificial Intelligence – A.I. capabilities or sophisticated control of their robotic artefacts. At this moment, trainers should be able to propose simple and stable techniques to their students for implementing such features in their constructions. This paper proposes flexible methods for this to be done by exploiting the very popular MIT App Inventor and Snap! visual programming environments, in conjunction with a modified tiny web server module, written in Python, that runs on a Raspberry Pi credit card - sized computer. Furthermore, this paper reports on simple techniques being used to make robust enough robots by low cost everyday/recyclable materials like cardboard, wood, plastic bottles or broken toys.

Keywords: DIY robots · Remote control · Visual programming tools · AI · App Inventor · Snap! · Raspberry Pi

1 Introduction

The fast growth of cell phone and tablet industry made possible the production of high-end electronics in vast numbers, and thus at very affordable costs. "Exotic" devices with features like wireless connectivity, cameras, GPS, motion sensors, high resolution

M. Moro et al. (Eds.): Edurobotics 2018, AISC 946, pp. 175–189, 2020.
https://doi.org/10.1007/978-3-030-18141-3_14

screens are offered at price levels that seemed unbelievable just few years ago. This fact gave a boost to the computer industry that started producing credit card - sized devices equipped with powerful features. Indeed, these devices arse much like cell phones without their cover and provide bigger General Purpose Input Output (GPIO) pins and other connectivity ports. Such devices, of diverse capabilities, ranging from Arduino [4] units to Raspberry Pi [19] ones, can be used in many embedded systems or IoT projects.

The benefits that similar systems might have in education weren't left unexploited [8]. Indeed, during the last years, we are witnessing a very successful osmosis between innovative and cost-effective credit card - sized computers and education. These computers, equipped with low cost sensors or actuators, can be the "heart" of various DIY robotic artefacts made by low cost everyday/recyclable materials like cardboard, plastic bottles or broken toys. This environment allows for a mixture of thinking and making activities that can be very meaningful in terms of pedagogy and science [1, 2]. Indeed, similar practices, usually referred as STEM or STEAM activities, are applied in many educational institutions, from primary schools up to universities, with most of the effort to focus on students of secondary education age.

Being in line with this movement, the eCraft2Learn EU H2020 [10] research project (duration: years 2017 and 2018) aims to introduce digital fabrication and maker movement in formal and informal education settings. The eCraft2Learn/H2020 research project researches, designs, pilots and validates an ecosystem based on digital fabrication and making technologies for creating computer-supported artefacts. The project aims at reinforcing learning and teaching in science, technology, engineering, arts and math (STEAM) education and the development of 21st century skills that promote inclusion and employability for youth in the EU. The eCraft2Learn ecosystem aspires to support both formal and informal learning.

In the frame of the eCraft2Learn research project a learning ecosystem is being designed offering technologies such as robotics, DIY electronics, 3D modeling & printing and visual programming tools supported by a constructionist learning methodology [7, 17]. The eCraft2Learn ecosystem is used to support pilot activities with students targeting on how to make robots and other computer – supported artefacts from scratch [10, 21]. In the framework of the eCraft2learn project, local labs were established and run project pilots in Athens (Greece) and Joensuu (Finland) in formal and informal education settings. In addition to student pilots, the eCraft2Learn labs are used for training teachers who act later as coaches in student pilots.

The overall process, although promising at the beginning and providing more freedom than using commercial out-of-the box educational robotic products [8], is not always straightforward to keep up with. More specifically, as students get more experience, they develop a "hunger" for more complicated scenarios that usually demand features like remote interaction with simple Artificial Intelligence [5] capabilities or sophisticated control of their robotic artefacts. At this moment, trainers should be able to propose simple and stable techniques to their students for implementing such features in their constructions [9], which is not always easy [12]. Indeed, although there are many quite cost - effective radio link modules available for boards like the Arduino, it is not always straight forward for the inexperienced user to make the necessary wiring as these modules usually demand extra electronic components to be added [11]. Furthermore, ordinary tablet or cell phone users, are not familiar with radio links different than the trivial Wi-Fi (i.e., the IEEE 802.11 [26]) and are unwilling to spend extra money, in order to buy equipment (at least in pairs) to remotely control their robotic artefacts. But, even when using ordinary Wi-Fi modules, one can be far from being efficient. Indeed, connection methods through VNC (i.e., Virtual Network Computing [25]) although straightforward to use, require a large amount of computational power and bandwidth to run flawlessly, but this is not an optimal choice when creating autonomous robotic artefacts.

This paper proposes flexible methods for this remote interaction and control process to be done flawlessly, via the familiar Wi-Fi links, by exploiting the very popular MIT App Inventor [15] or the Snap! [22] visual programming environments, in conjunction with a modified tiny web server module, written in Python [18], that runs on a Raspberry Pi credit card - sized computer.

The crafting experience that has been gained during the eCraft2Learn project is also reported as good practices have been tested to make robust enough robots by low cost everyday/recyclable materials like cardboard, wood, plastic bottles or broken toys.

Although the eCraft2learn project has initially been used to support pilot activities with students aged from 13 to 17, the techniques being studied and refined during this project are applicable to students of universities as well, after small necessary adaptations. Such successful tests have initially been carried out with student-teachers who attend the post-graduate course STEM Education, co-organized by the University of Patras and the National Kapodistrian University of Athens [3], and lately with students of the Agricultural University of Athens (Laboratory of Agricultural Engineering).

The rest of the paper is organized so as to provide the architecture being followed to support a flawless and easy remote interaction/control of the robotic artefacts (in Sect. 2), to highlight useful implementation details (in Sect. 3), to present some characteristic examples (in Sect. 4) and, finally, to conclude on the overall process and give directions for future work (in Sect. 5).

2 Architectural Overview

One of the most crucial components of the eCraft2Learn ecosystem is the Raspberry Pi (model 3 B) single board computer (RPi3). This unit appears to have more than one important role.

Indeed, a set of Raspberry Pi 3 units are serving as experimental development computers (workstation units) for the students. These units are equipped with TFT screens and keyboard-mouse sets. The adoption of this solution tackles the problems related to the vast software and hardware diversity characterizing a typical School Lab and reduces the Lab's overall energy consumption and, unlike computers and tablets controlled by a school's IT staff, typically teachers may have full control over their Raspberry Pis. Furthermore, a second set of Raspberry Pi 3 units are used for the potential implementation of the designed artefacts (the more complex ones). Of course one or more Arduino boards, connected with a variety of electronic components and/or the Raspberry Pi 3 units, to assist in artefact implementation, are provided as well.

The overall ecosystem, in terms of hardware, includes tablet devices to facilitate remote interaction scenarios and various DIY electronic components (e.g. photoresistors, potentiometers, servomotors, LEDs) that will be used in conjunction with the RPi3 and the Arduino units to implement the designed artefacts and various DIY modified parts brought from home during a recycling process, like broken toys, plastic bottles, pieces of paperboard, computer fans, speakers, etc. Moreover, this ecosystem includes 2D and 3D printing equipment, the necessary networking equipment to facilitate the interconnection of the pilot sites and provide further access to the Internet and eCraft2Learn resources, and, finally the necessary power supply equipment like power banks (or small solar panels). The selection of all the hardware components has been done to minimize the cost, the size, the power consumption and maximize the reusability of materials and electronics.

The eCraft2Learn ecosystem, in terms of software, includes a set of tools of diverse capabilities that have been selected to meet a set of very welcome characteristics. More specifically, they can run on a RPi3 environment, they have reduced need of installation or update of software elements, have a user-friendly interface, are pedagogically meaningful, are easy to integrate with the external hardware, are open source and free (or at least have low cost) and, finally, are preferably capable of supporting Artificial Intelligence cloud services. This basic set of software tools is extensible with more tools dedicated to more advanced users.

Amongst the above mentioned software tools are visual programming tools like the Snap! (and its variants) that allows custom visual block creation, provides easily exportable/importable XML code and cooperates with many web-based tools for useful data exchange. As the vast majority of today's students are very familiar with smart phones/tablets, another outstanding tool is the MIT App Inventor that allows for rapid mobile application creation even by inexperienced users.

As students are getting more and more familiar with the capabilities and strengths of the abovementioned ecosystem they usually ask to create more sophisticated artefacts and tend to move to more composite scenarios of human – robot interaction. At this point, both MIT App Inventor and Snap! can provide satisfactory solutions, in conjunction with the RPi3 units and a few lines of code written in Python (that might remain hidden from the final users or not), without increasing complexity or relying upon hiring "exotic" radio interfaces. Furthermore, the utilization of a quite powerful unit (like the Raspberry Pi) inside the robotic artefacts allows for more composite human – robot interaction.

The key factor to achieve that is to take advantage of the HTTP request/reply messaging mechanism that most today's systems are supporting as a common feature. More specifically, both MIT App Inventor and Snap! support such a mechanism and, in order the RPi3 unit to be able to intercept and handle HTTP requests, Python implemented HTTP server code has been modified appropriately and adjustments have been made so as this Python code to be executed as a service on the RPi3 unit (i.e., to start automatically after powering up the RPi3 unit).

Snap! and many of its variants such as Snap4Arduino, NetsBlox, and BeetleBlocks can read the contents of URLs. New blocks can be defined to behave just like the Snap4Arduino [23] blocks that read and write digital pins of Arduinos. For example the Raspberry Pi's digital pin 5 can be written by a block that can write any pin by constructing URLs such as localhost/writePin/5/on. The small Python web server then turns on the pin. To read a pin a URL such as localhost/readPin/5 is generated and depending upon whether the contents of the URL are 1 or 0 will report the appropriate Boolean value.

Instead of localhost one can use raspberrypi.local instead. The Raspbian operating system treats this the same way as localhost, when fetching pages from the Raspberry Pi. But it also enables access to the Python-implemented web server from any device on the local area network. This means, for example, that a laptop on the local network can read and write the Raspberry Pi pins using the Snap! blocks via Wi-Fi. This enables programming the Raspberry Pi without connecting a display, keyboard, and mouse to it. Furthermore, if the project involves machine learning it can rely upon the browser's access to the GPU to speed up computations over one hundred times over the speed of the Raspberry Pi.

Figure 1 depicts the interoperation among the various modules supporting the robot control actions.

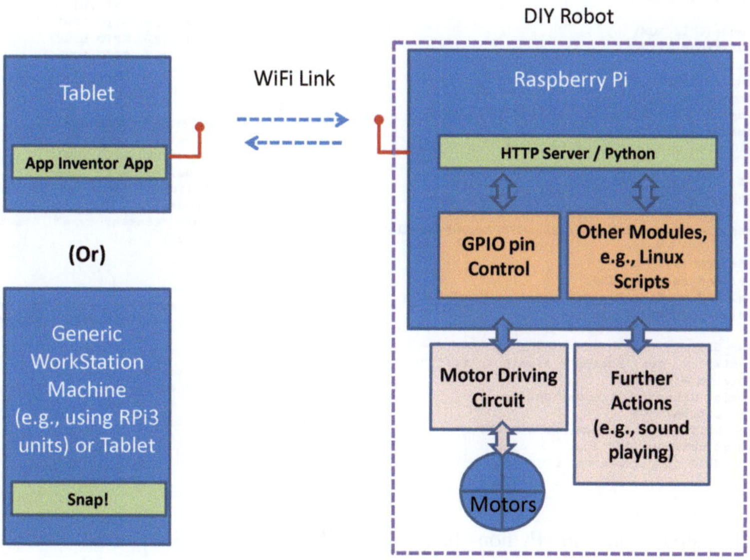

Fig. 1. Diverse modules interoperation to support the robot control functions.

3 Implementation Details

In further detail, a simple HTTP server code like the one described in [6] has been combined with the RPi.GPIO Python library [20] so as to control the GPIO pins of the RPi3 board. This Python related part, most of which may remain transparent for the learners, provides the ability to handle the necessary HTTP requests (of GET, PUT or POST type) into real (remote) interaction cases. Figure 2 depicts indicative code (in Python) that controls the status of GPIO pins of the RPi3 unit (e.g., by handling the http://127.0.0.1:42101/GPIO_18_ON request). The Python code includes separate cases that can be further extended so as instead of simply altering the status of some GPIO pins to call custom Linux (bash) scripts performing even more sophisticated actions like sound/music invocation or battery status data acquisition. Indeed by importing os [16] library in python code we can invoke any system command (or script) to be executed via the os.system() method. For instance by including the "os. system("mpg321 ScorpionsWind.mp3")" command in our python code we invoke the song ScorpionsWind.mp3 to start playing (under a specially designed remote command of course).

This Python code can be set up to run as a service that start automatically each time the RPi3 units is powered on, by adding an entry in the /etc./rc.local file and/or the Linux crontab.

```
import time
import BaseHTTPServer
import cgi
import RPi.GPIO as GPIO

GPIO.setmode(GPIO.BCM)
GPIO.setwarnings(False)
GPIO.cleanup()
GPIO.setup(17, GPIO.OUT)
GPIO.setup(18, GPIO.OUT)
GPIO.setup(22, GPIO.OUT)
GPIO.setup(23, GPIO.OUT)
#more GPIO pins can be configured here ...

HOST_NAME = '0.0.0.0'
PORT_NUMBER = 42101

class MyHandler(BaseHTTPServer.BaseHTTPRequestHandler):

    def do_HEAD(s):
        s.send_response(200)
        s.send_header("Content-type", "text/html")
        s.end_headers()

    def do_GET(s):
        category, pin, action = s.path.split("_");
        pinnum = int(pin)
        print category, pinnum, action
        if action=='ON':
          GPIO.output(pinnum, GPIO.HIGH)
        elif action=='OFF':
          GPIO.output(pinnum, GPIO.LOW)
          ...
```

```
...
if action=='FD':
            GPIO.output(17, GPIO.HIGH)
            GPIO.output(18, GPIO.LOW)
            GPIO.output(22, GPIO.HIGH)
            GPIO.output(23, GPIO.LOW)
elif action=='BK':
  ...
elif action=='ST':
            GPIO.output(17, GPIO.LOW)
            GPIO.output(18, GPIO.LOW)
            GPIO.output(22, GPIO.LOW)
            GPIO.output(23, GPIO.LOW)
        #response2client
        s.send_response(200)
        s.send_header("Content-type", "text/html")
        s.end_headers()
        return

if __name__ == '__main__':
    server_class = BaseHTTPServer.HTTPServer
    httpd = server_class((HOST_NAME, PORT_NUMBER),
MyHandler)
    try:
        http.serve_forever()
    except KeyboardInterrupt:
        pass
    httpd.server_close()
```

Fig. 2. Indicative code in Python handling HTTP (GET) requests and activating the corresponding GPIO pins on the Raspberry Pi unit of the robot.

The eCraft2Learn learners were now able to create simple HTTP requests, allowing for direct access/control of the GPIO pins of either local or remote RPi3 units. For this to be done, the HTTP block functionality in Snap!, that is available under the "Sensing" category via the "url …" block, was necessary. By forming suitable HTTP requests, a LED can be turned on and off or a robot commanded to move back and forth, as depicted in Fig. 3.

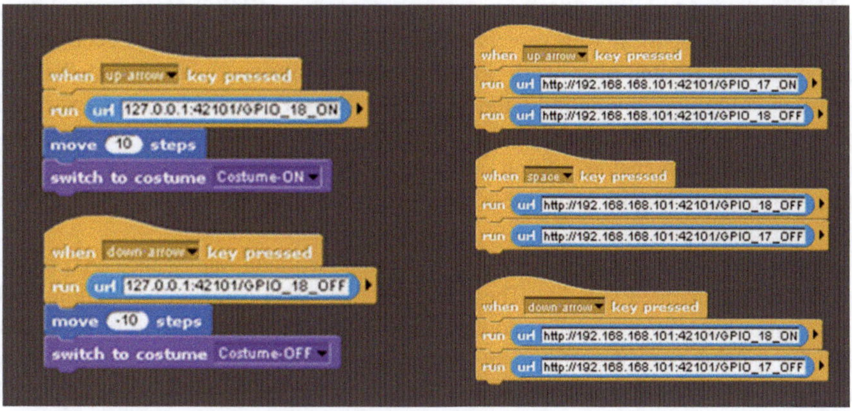

Fig. 3. Suitable HTTP requests in Snap! to turn a LED on and off (left) or to command a robot to move back and forth (right).

Multiplayer games and projects can be created by arranging for multiple Snap! instances to communicate with the same Raspberry Pi. This relies upon the fact that the Raspbian operating system supports the detection of Raspberry Pis on the local area network unless default settings have been overridden.

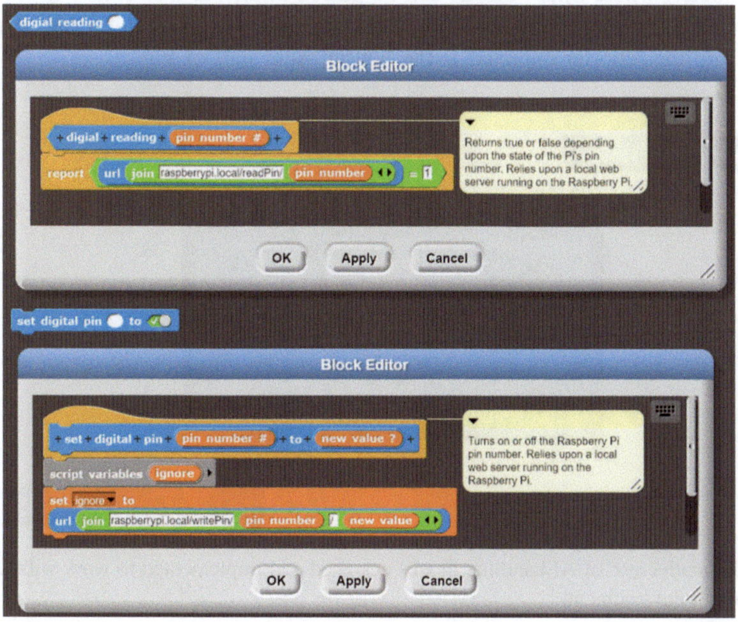

Fig. 4. The definition of blocks in Snap! compatible with Snap4Arduino's digital pin blocks.

Unlike Snap4Arduino that requires access to the serial port connected to an Arduino the use of local web servers means that the Snap! blocks for accessing the Raspberry Pi's digital pins work in all modern browsers without installing any extensions. Moreover, by defining the Snap! Raspberry Pi blocks to match exactly the digital pin blocks of the Snap4Arduino, projects can be developed so as to work equally well without custom editing on Arduinos or Raspberry Pis. Figure 4 shows an implementation of blocks compatible with Snap4Arduino (it uses a slightly different local web server).

During the eCraft2Learn project very important Artificial Intelligence (AI) features have been implemented [13, 14]. These features like voice command reception can be easily combined with the remote control capabilities described here to provide pedagogically fruitful examples of human – robot interaction. This relies upon the speech recognition API implemented in the Chromium browser. Image recognition can also be performed in Snap! from the browser if the appropriate API keys are provided.

Snap! machine learning blocks can run on a Raspberry Pi but is typically slow. However, more powerful computers can use machine learning and control the pins of a Raspberry Pi using the raspberrypi.local domain feature.

The AI cloud service Snap! blocks range from very simple interfaces to fully functional ones with many options. For a majority of projects the simple interfaces are adequate. For example, speech recognition can be as simple as using the "the next thing spoken" reporter (Fig. 5 at top left.) which reports a text string for the next thing spoken. For more complex applications one can use a specially designed block (Fig. 5 at the middle.) where Snap! blocks can be placed in any of the "rings" to receive the appropriate inputs. Similarly for image recognition one can obtain a list of possible labels using the reporter depicted at the bottom right part of Fig. 5.

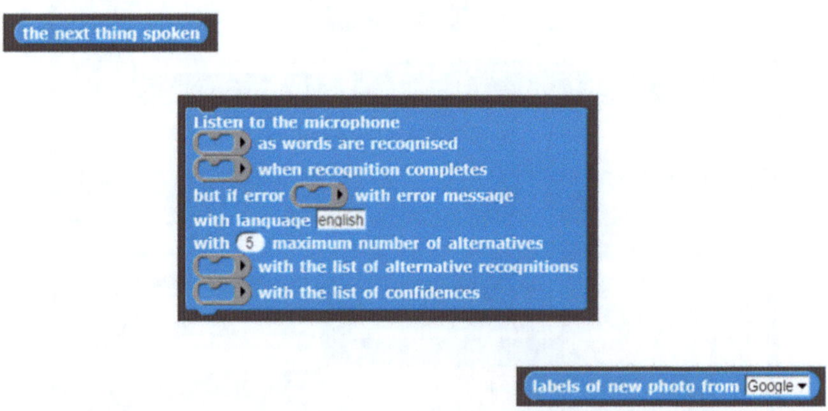

Fig. 5. Indicative use of AI handling blocks designed and implemented to work with the Snap! programming environment.

The architecture we developed splits responsibilities in order for a Snap! program to access both AI cloud services and the sensors and actuators connected to the

Raspberry Pi's pins. The new Snap! blocks we authored in JavaScript make HTTP(S) requests to both the local web server implemented in Python and to cloud services hosted by companies such as Google, Microsoft, and IBM. The local web server is only used to access the pins of the Raspberry Pi. The blocks communicating with cloud services can access the camera and microphone from the browser in the standard manner that any webpage does so. Typically the first time permission must be granted. The blocks then capture images and audio and send them to the cloud service with HTTPS POSTs.

The Snap! AI blocks are described in more detail in [13]. They have been the focus of AI programming workshops in six countries involving over one hundred students ranging from 8 to 17 years old. The children 8 to 12 years old successfully used the AI blocks due to their prior experience with Scratch programming. Some of the high school students succeeded despite little or no prior programming experience.

Our architecture relies upon a typical networking setup including a router/access point device to maintain a wireless LAN (containing both the tablets and the RPi3s) and to provide access to the Internet (i.e., to the Cloud services) via a WAN line (e.g., a DSL line). In case that tablets are equipped with their own 3G/4G interface, typical tethering techniques can be used to provide access to both the RPi3 unit and the Internet, and thus no extra router is needed. Typically, the single wireless network interface card, which is standard in the RPi3 units, is sufficient. The HTTP(S) requests to the local web server are handled by the Raspbian operating system, without using any dedicated network hardware, via the loopback interface (i.e., using the localhost address which is by default the 127.0.0.1). In case that Snap! runs in a different device (either a tablet or a RPi3 unit) than the RPi3 unit having the GPIO pins of interest, the localhost address should point at the IP address of this specific RPi3 unit. Apparently, more complicated networking schemas can be applied but such configurations do not further highlight the objectives of this research work, and thus are omitted.

The students can use the Snap! blocks for reading and writing the pins of a Raspberry Pi without understanding how they work by making requests to the local web server. However, the idea that a URL that indicates that a pin should be read will respond with the state of the pin and another URL for writing a pin will turn on or off a device connected to the specified pin is not difficult for the students to understand as modern students, even the younger ones, are quite familiar with web environments thus to their everyday social media and gaming activities.

The simple custom mechanism that was added to the RPi3 unit to handle HTTP requests is beneficial not only for the Snap! environment but for any other tool able to generate HTTP requests in a pedagogically meaningful manner. So, as a next step, tasks similar to the above ones were practiced in the MIT App Inventor environment. More specifically, the learners generated HTTP requests in order to remotely control GPIO pins of their RPi3 unit, through their smart phones or tablets.

They were then invited to experiment with the Accelerometer sensor which is built in any modern smart phone in order to turn on and off the GPIO18 pin of the local RPi3 or the same pin of the RPi3 unit of their neighboring team or to drive a robot as they did via the Snap! software. Figure 6 depicts an indicative user interface layout and code part (written in MIT App Inventor) that controls two DC motors connected on pins 17,18 and 21,22 of the RPi3 respectively (using the necessary driving circuit of course, like the L293D one). The typical robotic artefact construction being made has one independent motor per side. At the bottom of Fig. 6, code that controls one separate GPIO pin is presented.

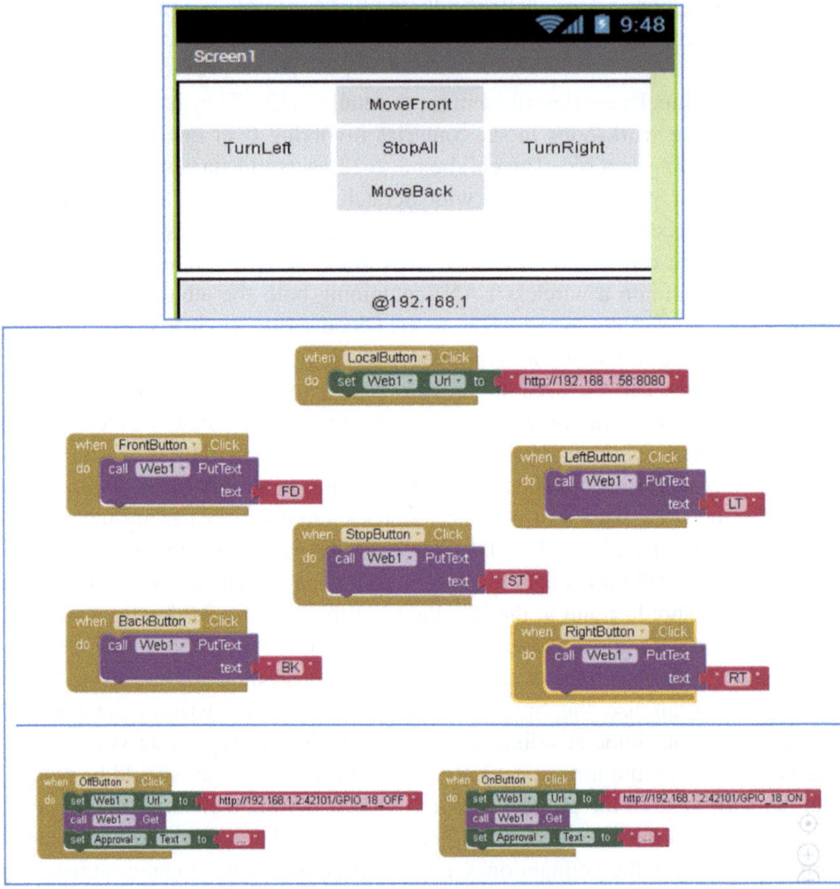

Fig. 6. Indicative user interface layout and code, in MIT app inventor, sending the suitable HTTP requests each time the corresponding virtual button is pressed on the smart phone's screen.

To further highlight how the above mentioned techniques can be combined to control a robotic artifact, exactly this task was implemented twice, using both MIT app inventor and Snap!. For simplicity reasons, in this case, the robotic artefact was designed to have only one motor to move it, and thus only two GPIO pins are required to drive it. Most experiments implemented with the assistance of students of the Agricultural Engineering laboratory of the Agricultural University of Athens. The left part of Fig. 7 depicts indicative code, in MIT app inventor, that intercepts simple voice commands (after shaking the smart phone or touching a specific button on it). The right part of Fig. 7 depicts the Snap! counterpart of the same process. More specifically, user can order the robot to move in front, to stop, or to go back by saying the suitable command. In both cases, the program interprets the voice commands in suitable ON/OFF action requests, through forming the suitable HTTP messages that involve the GPIO pins 17 and 18 of the Raspberry Pi unit. These action requests are intercepted by the web server running on the Raspberry Pi and the proper pins alter their state properly, through suitable RPi.GPIO specific commands.

It must be noted that the speech recognition, incorporated into the robotic artefact projects, is a time consuming process requiring at about 2–3 s to be accomplished (in terms of both networking and processing costs). This non negligible delay has to be taken into account when designing human – robot interaction scenarios, in order the later to be efficient and user friendly.

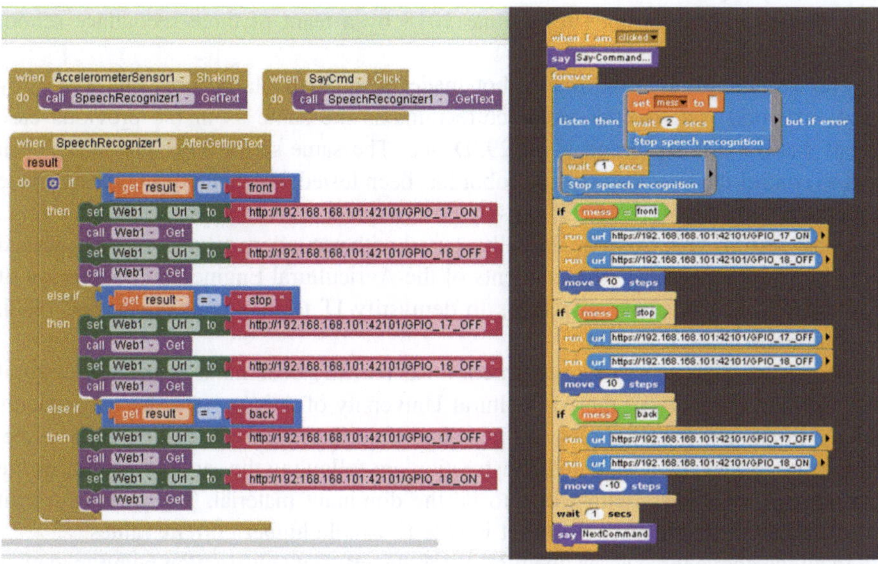

Fig. 7. Indicative code in MIT app inventor (left) and Snap! (right) intercepting simple voice commands from user and sending the suitable HTTP requests towards the robot.

Potential bottlenecks to the above mentioned techniques may rise due to the security settings that some browsers have eliminating access over unencrypted HTTP messages in favor of HTTPS ones. This is an apparent problem, especially in case of Snap!, and can be solved by replacing the HTTP server in Python with an equivalent HTTPS version. The necessary security certificates should be provided then in the Raspberry Pi unit inside the robotic artefact. An alternative solution might be to implement pure UDP (i.e., User Datagram Protocol [24]) code for client/server interoperation, instead of HTTP, but this goes beyond the scope of this paper, as it requires deeper networking experience by the students being involved and more sophisticated code implementation.

4 Characteristic Examples

This section is dedicated to presenting some characteristic robotic artefacts made by the eCraft2Learn learners (both secondary school students and university students), that are low cost, easy to make and combine simplicity and robustness in their construction and operation.

The 1st example is a cardboard robot having separately driven wheels by small DC motors that are glued at each side, plus a free wheel. The communication and control is carried out by a RPi3 unit and the built in Wi-Fi module. Pins 17, 18 are controlling the left side motor while GPIO pins 21, 22 the right side motor. A pair of L9110S motor control chips (on the same board) is used to amplify the signals of the GPIO pins to efficiently drive the motors. The whole system is powered by a solar power bank. This robot was easily constructed along with is client software, made using the App Inventor tool, in only two sessions of two hours each. The activities took place in Athens eCraft2Learn informal pilot site in June 2018 by a team of three secondary school students.

The 2ndexample is a minimal robot made by a small plastic box and some tiny pieces of wood. The overall architecture remains the same as in the previous case except that the driving circuit is a L293D one. The same set of RPi3 pins have been used to control the two motors. The robot has been tested using an App Inventor made mobile app and separate HTTP messages through a conventional web client (e.g., Mozilla Firefox). This example, initially started with secondary school students, served as an introductory project so as students of the Agricultural Engineering laboratory of the Agricultural University of Athens to demystify IT technologies and visual block programming.

The 3rd example of robotic construction has recently been made at the Agricultural Engineering laboratory of the Agricultural University of Athens, in order the students become familiar with techniques and machines used in modern agriculture under scale. Apart from its size, its design and construction are following the same principles as the two previous examples, with wood to be the dominant material. The driving circuit now is based on the L298N chip that is able to handle higher current values.

In all the above three cases, the typical robotic artefact construction being made has one independently controlled motor per side. This setup eliminates the need for extra mechanisms dedicated in steering tasks and leads to more robust artefacts that are able to maneuver/turn in narrow areas just by changing the rotation direction of their side wheels.

Learners, supported by their trainer, worked in high interest, realized successfully and enjoyed their projects. Team members had discrete and diverse roles (e.g., crafting, programming, presentation) and joined their powers to accomplish the project.

Figure 8 depicts these three characteristic DIY robot cases: Bottom left picture corresponds to the 1st example. Top middle corresponds to the 2nd example. Bottom right picture corresponds to the 3rd example. Despite the noticeable difference in size, all three implementations are sharing the same architecture, in terms of hardware and software, and thus they can be controlled by the same application interface.

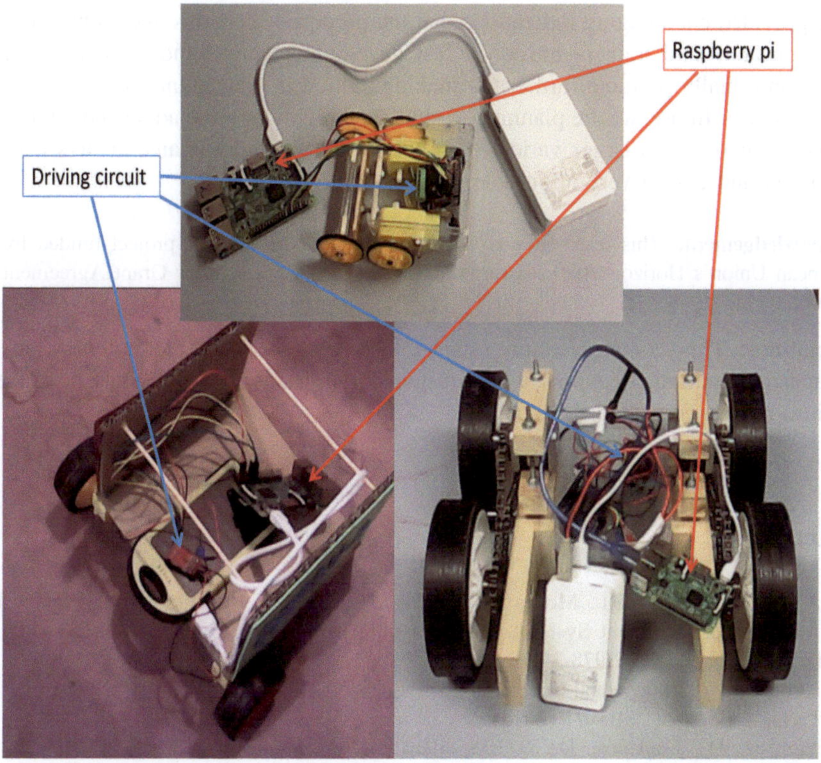

Fig. 8. Examples of DIY robots of diverse size that are all sharing the same architecture and exhibit cost effectiveness and ease to use and handle.

5 Conclusions and Future Work

This paper proposed flexible methods for remotely controlling and interacting with DIY robotic artefacts by exploiting the very popular MIT App Inventor or the Snap! visual programming environment, in conjunction with a modified tiny web server module, written in Python, that runs on a Raspberry Pi credit card - sized computer. The proposed techniques provide solutions in situations where students involved in STEM/STEAM education activities, as they get more experience, want to realize more complicated scenarios that usually demand features like remote interaction with or sophisticated control of their robots.

This research work exploited the experiences gained during the eCraft2Learn EU research project that aims to introduce digital fabrication and maker movement in formal and informal education settings, and reported on simple techniques being used to make robust enough robots by low cost everyday/recyclable materials like card-board, wood, plastic bottles or broken toys. The techniques being presented, although initially planned for secondary school students have successfully been tested with university students as well. The experiences gained from these first tests with learners have provided encouraging indications that the proposed scenarios and techniques can help educators and trainers introduce fast development and remote control of DIY robots and finally promote further the making culture in educational robotics.

In the near future, we are planning to study and highlight more advanced interaction scenarios involving a larger variety of robotic artefacts and humans, always keeping feasibility and cost at very affordable level.

Acknowledgement. This research was supported by the eCraft2Learn project funded by the European Union's Horizon 2020 Research and Innovation Action under Grant Agreement No 731345.

Disclaimer: This communication reflects the views only of the authors and the European Commission cannot be held responsible for any use which may be made of the information contained therein.

References

1. Alimisis, D., Moro, M., Menegatti, E. (eds.): Educational Robotics in the Makers Era. Advances in Intelligent Systems and Computing, vol. 560. Springer, Heidelberg (2017). https://doi.org/10.1007/978-3-319-55553-9
2. Alimisis, D.: Educational robotics: open questions and new challenges. Themes Sci. Technol. Educ. **6**(1), 63–71 (2013)
3. Alimisis, D., Loukatos, D.: STEM education post-graduate students' training in the eCraft2Learn ecosystem. In: 2nd International Conference on Innovating STEM Education, Athens (2018, in press)
4. Arduino. https://www.arduino.cc/. Accessed May 2018
5. Artificial Intelligence – AI. https://en.wikipedia.org/wiki/Artificial_intelligence. Accessed May 2018
6. BaseHTTPServer. https://wiki.Python.org/moin/BaseHttpServer. Accessed May 2018
7. Blikstein, P.: Digital fabrication and 'making' in education: the democratization of invention. In: Walter-Herrmann, J., Büching, C. (eds.) FabLabs: Of Machines, Makers and Inventors. Transcript Publishers, Bielefeld (2013)
8. Doran, M.V, Clark, G.W.: Enhancing robotic experiences throughout the computing curriculum. In: SIGCSE 2018, Baltimore, MD, USA, 21–24 February 2018, pp. 368–371 (2018)
9. Eaton, E.: Teaching integrated AI through interdisciplinary project-driven courses. AI Mag. **38**(2), 13–21 (2017)
10. eCraft2Learn EU H2020 project. https://project.ecraft2learn.eu. Accessed May 2018
11. ESP8266 custom shield for arduino. https://www.instructables.com/id/Cheap-Arduino-WiFi-Shield-With-ESP8266/. Accessed Sept 2018
12. Ferrarelli, P., Lázaro, M.T., Iocchi, L.: Design of robot teaching assistants through multi-modal human-robot interactions. In: Lepuschitz, W., Merdan, M., Koppensteiner, G., Balogh, R., Obdržálek, D. (eds.) RiE 2017. AISC, vol. 630, pp. 274–286. Springer, Cham (2018). https://doi.org/10.1007/978-3-319-62875-2_25
13. Kahn, K., Winters, N.: AI programming by children. In: Constructionism Conference, Vilnius, Lithuania, August 2018 (to appear)

14. Kahn, K., Winters, N.: Child-friendly programming interfaces to AI cloud services. In: Lavoué, É., Drachsler, H., Verbert, K., Broisin, J., Pérez-Sanagustín, M. (eds.) EC-TEL 2017. LNCS, vol. 10474, pp. 566–570. Springer, Cham (2017). https://doi.org/10.1007/978-3-319-66610-5_64
15. MIT App Inventor. http://appinventor.mit.edu/explore/. Accessed May 2018
16. OS - Miscellaneous operating system interfaces (Python). https://docs.python.org/3/library/os.html. Accessed May 2018
17. Papert, S., Harel, I.: Preface, situating constructionism. In: Harel, I., Papert, S. (eds.) Constructionism, Research Reports and Essays, 1985–1990, p. 1. Ablex Publishing Corporation, Norwood (1991)
18. Python Programming Language. https://en.wikipedia.org/wiki/Python_(programming_language). Accessed June 2018
19. Raspberry. https://www.raspberrypi.org/. Accessed May 2018
20. RPi.GPIO module basics (Python). https://sourceforge.net/p/raspberry-gpio-python/wiki/BasicUsage/. Accessed June 2018
21. Schon, S., Ebner, M., Kumar, S.: The maker movement implications from modern fabrication, new digital gadgets, and hacking for creative learning and teaching. In: Canals, L. (ed.) eLearning Papers Special edition 2014, pp. 86–100 (2014). http://www.openeducationeuropa.eu/en/article/Learning-in-cyber-physical-worlds_In-depth_39_2
22. Snap! https://snap.berkeley.edu/. Accessed May 2018
23. Snap4Arduino. http://snap4arduino.rocks/. Accessed May 2018
24. UDP – User Datagram Protocol. https://en.wikipedia.org/wiki/User_Datagram_Protocol. Accessed June 2018
25. VNC – Virtual Network Computing. https://en.wikipedia.org/wiki/Virtual_Network_Computing. Accessed June 2018
26. Wi-Fi – IEEE 802.11 Standard. http://www.ieee802.org/11/. Accessed May 2018

Short Papers Reporting Good Practices or Work in Progress (Presented in the Conference as Posters)

How to Support Girls' Participation at Projects in Makerspace Settings. Overview on Current Recommendations

Sandra Schön[1(✉)], Margarethe Rosenova[1], Martin Ebner[2], and Maria Grandl[2]

[1] Salzburg Research Forschungsgesellschaft, Jakob Haringer Strasse 5/III, 5020 Salzburg, Austria
sandra.schoen@salzburgresearch.at
[2] Graz University of Technology, Münzgrabenstraße 36, 8010 Graz, Austria

Abstract. Several biases and thresholds challenge the reach of girls in technology-related activities. For this contribution we collected and structured existing research and good practices on how to reach girls within projects in the field educational robotics, makerspaces, coding and STEM in general. The contribution presents general guidelines for future activities with a potential higher rate of participating girls in makerspace settings.

Keywords: Girls · Participation maker education · STEM · Makerspace · Robotics

1 Introduction

Several biases challenge the work with girls, e.g.: teachers have perceptions that boys are more interested in technology [1] or that makerspaces are not safe for girls [2]. Catalonian boys from 11 to 13 reach higher self-efficacy for doing tasks with computers than girls [3]. Already in the age of kindergarten children, they begin to decide "which technology and engineering activities and materials are better suited for boys or girls" [4]. These biases - and other framework conditions - result in the fact, that girls are typically underrepresented in activities from the field of educational robotics, makerspaces and coding and that women are underrepresented amongst engineers, scientists, IT experts and related domains. There are a wide variety of approaches to influence this and to gain higher share of females, which is amongst others one of the sustainable development goals of the UNESCO [5]. For the authors and their work fields it is important to reach girls within their maker activities: The Horizon 2020 project "DOIT - Entrepreneurial skills for young social innovators in an open digital world" (2017–2020) - co-financed by the European Union - builds upon the consideration that social innovations in makerspace settings allow authentic learning experiences fostering future entrepreneurial spirit and ambition to co-create a (better) world (see http://DOIT-Europe.net, [6]). Graz University of Technology (TU Graz) as well takes socio-political aims seriously and therefore highlights and facilitates diversity (cf. Office for Gender Equality and Equal Opportunity at TU Graz).

M. Moro et al. (Eds.): Edurobotics 2018, AISC 946, pp. 193–196, 2020.
https://doi.org/10.1007/978-3-030-18141-3_15

2 Research Issue

For our future activities we looked for relevant literature that gives us advice on how we can reach (more) girls within our activities within makerspaces as they are currently underrepresented. This includes answers on the following sub-questions: What do others do to reach girls? What do they recommend? Therefore we collected existing experiences from projects and research (including literature on girls and maker education, girls in makerspaces, girls and robotics from the last five years in the ERIC database; 2013–2018).

3 Recommendations from Literature and Projects

3.1 Approach and Overview

For the condensed recommendations we used recommendations building upon research (e.g. [7]) as well as the experiences collected in the Gender Action Guidelines of the EU project Phalabs 4.0 [8]. Figure 1 gives an overview about stages within the project/activity development that need a special awareness concerning gender.

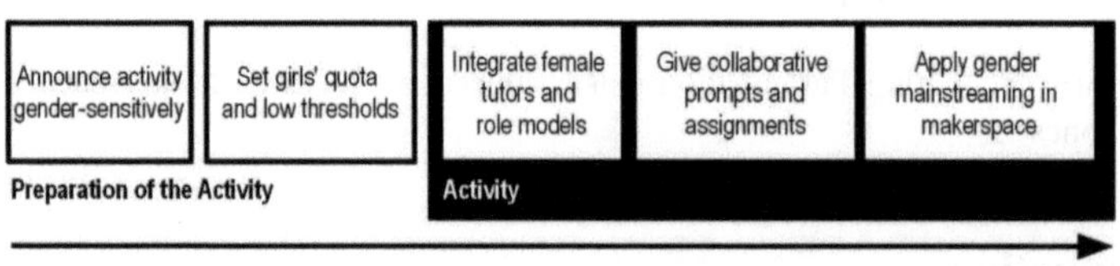

Fig. 1. Overview: guidelines to reach girls in makerspace settings

3.2 The Guidelines

The following is a condensed description with further references to the sources.

Gender-Sensitive Announcement of the Activity. Girls tend to get (more) motivated if a title of the event or activity includes not only its contents, e.g. "Robotics with kids", but to get a sense of value of the activity, e.g. "Robotics for gardeners" or "Robotics within the book sector". It is recommended to highlight the value of the activity already in the title, e.g. the impact to the world [8]. When marketing a measure, it is also helpful to ensure that it is not advertised to (future) engineers, scientists, and mathematicians. There is evidence to suggest that such professional identities are less common among girls [9], especially, if they are from minorities, and can therefore be less appealing. Gender-sensitive language and gender-sensitive illustration are known as important to girls [8] which includes e.g. that girls are shown as active participants in the marketing materials.

Set Girls' Quota and Low Thresholds. If it is planned that children have to be registered for an event, the proportion of girls may be smaller - at least the participation of boys in technology-related offers is rather supported by (grand) parents, as experience at the Maker Days has shown [11]. On the other hand, an enrolment procedure and confirmation of enrolment also allows a quota to be set for girls. Macdonald [8] gives the advice to always insist on the 50:50 rule, if mixed schools are partnering (p. 8).

Female Tutors and Role Models. A same-sex role model seems to be a strong supporter to help girls to get in touch with technology. The "Maker Days for Kids" for example was a creative digital workshop that was open for four days in April 2015 for children aged 10 to 14 where 44% of the participants were girls (no registration, no selection, no fees, N = 69) [11, 12]. All participants could decide what to do in the open makerspace, including to participate in short workshops, e.g. in the devlab area where one female tutor and three male tutors were active. Girls chose more workshops offered by female tutors than boys (31 vs. 24%, 189 workshop participations in devlab, p. 98) [12]. Role models are great, especially if they can furthermore tell stories about how they failed or other personal stories [8].

Collaborative Prompts and Assignments. Girls prefer activities that are collaborative. Collaborative assignments lead to a higher future interest in STEM-related fields [7, p. 124). Females will prefer activities that are collaborative, which means that there is a positive outcome for all that are involved - and that are no competitions [8]. Practical examples are available [10, 11, 14].

Gender Mainstreaming in Project Activities. Practically, teachers or tutors can contribute to gender disparities in makerspace settings as well [1]. Active gender mainstreaming in maker activities could therefore include gender mainstreaming along the whole activity: "If supervising a school group, ensure you spend as much time talking to the females as talking to the males. Males often demand more of teacher's attention (often by doing silly things) while females get on with the task in hand. Females then perceive that they are of less value in STEM as teacher didn't talk to them very much or ask how they were getting on" [8]. Gender mainstreaming in project activities includes e.g. considerations such as give girls the same attention as boys, girls should be similarly participating e.g. at presentations of group work. It should be noted here that such a conscious - but not compulsive - proposal does not necessarily meet with public approval [13].

4 Discussion and Open Issues

There is still a need to consolidate, reflect and share experiences and results on how to deal with girls' participation. This analysis e.g. ignored diverse backgrounds of culture and educational systems of studies and literature.

Within our own projects we aim to get deeper insights into how e.g. design decisions and the availability of female role models will influence girls' participation and will establish fitting research. In DOIT therefore an evaluation of activities with about 1.000 children in 10 regions all over Europe is planned and available in future. Furthermore we will promote the girls' issues in maker education [15].

References

1. Legewie, J., DiPrete, T.A.: The high school environment and the gender gap in science and engineering. Sociol. Educ. **87**(4), 259–280 (2014)
2. Wittemyer, R., et al.: MakeHers: engaging girls and women in technology through making, creating, and inventing. Study by Intel Corporation (2014)
3. Cussó-Calabuig, R., Carrera Farran, X., Bosch-Capblanch, X.: Are boys and girls still digitally differentiated? The case of Catalonian teenagers. J. Inf. Technol. Educ. Res. **16**, 411–435 (2017)
4. Sullivan, A., Bers, M.U.: Girls, boys, and bots: gender differences in young children's performance on robotics and programming tasks. J. Inf. Technol. Educ. Innov. Pract. **15**, 145–165 (2016). http://www.informingscience.org/Publications/3547. Accessed 15 July 2018
5. UNESCO: STEM and gender advancement (SAGA) (project homepage). https://en.unesco.org/saga. Accessed 14 July 2018
6. Schön, S., Jagrikova, R., Voigt, C.: Social innovations within makerspace settings for early entrepreneurial education - the DOIT project. In: Proceedings of the EdMedia Conference, Amsterdam, 25–29 June 2018, pp. 1716–1725. (2018). http://www.learntechlib.org/primary/j/EDMEDIA/v/2018/n/1/. Accessed 15 July 2018
7. Hyun, T.: Middle school girls: perceptions and experiences with robotics. ProQuest LLC, Ed.D. dissertation, California State University, Fullerton (2014). https://search.proquest.com/docview/1553689179. Accessed 15 July 2018
8. Macdonald, A.: Gender action guidelines. Inventory of considerations that must be taken into account during the development of a workshop for Fab Labs. Phalabs 4.0 - PHotonics enhanced fAB LABS supporting the next revolution in digitalization (2018). http://www.phablabs.eu/. Accessed 15 July 2018
9. Tan, E., Calabrese Barton, A., Kang, H., O'Neill, T.: Desiring a career in STEM-related fields: how middle school girls articulate and negotiate identities-in-practice in science. J. Res. Sci. Teach. **50**(10), 1143–1179 (2013)
10. Kekelis, L., Ryoo, J., McLeod, E.: Making and mentors: what it takes to make them better together. Afterschool Matters **26**, 8–17 (2017)
11. Schön, S., Ebner, M., Reip, I.: Kreative digitale Arbeit mit Kindern in einer viertägigen offenen Werkstatt. Medienimpulse, no. 1 (2016). https://www.medienimpulse.at/articles/view/829. Accessed 15 July 2018
12. Gappmaier, L.: MakerDays for kids – analyse und Konzepterstellung. In: Ebner, M., Schön, S. (eds.) iTUG, vol. 8. Book On Demand. Norderstedt (2018). https://itug.eu. Accessed 23 July 2018
13. Kuhar, R., Zobec, A.: The anti-gender movement in Europe and the educational process in public schools. CEPS J. **7**(2), 29–46 (2017)
14. Gomoll, A., Hmelo-Silver, C., Šabanović, S., Francisco, M.: Dragons, ladybugs, and softballs: girls' STEM engagement with human-centered robotics. J. Sci. Educ. Technol. **25**(6), 899–914 (2016)
15. Schön, S., Ebner, M., Kumar, S.: The maker movement. Implications of new digital gadgets, fabrication tools and spaces for creative learning and teaching. In: eLearning Papers, vol. 39, pp. 14–25 (2014)

Skills Behind the Robotics – How to Re-educate Workers for the Future

Hanna Nygren[1(✉)], Kari Nissinen[1], Juhani Rautopuro[1], Kati Mäkitalo[2], and Kari Ullakko[3]

[1] Finnish Institute for Educational Research, University of Jyväskylä, Jyväskylän, Finland
{hanna.nygren, juhani.rautopuro}@jyu.fi
[2] University of Oulu, Oulu, Finland
kati.makitalo@oulu.fi
[3] Lappeenranta University of Technology, Lappeenranta, Finland
kari.ullakko@lut.fi

Abstract. The aim of this study is to respond to the educational needs of the future, considering automation and robotics. It is inevitable that automation and robotics are changing our lives and they create challenges for the future work life and education. In this study, we investigate what is the educational background of the unemployed people who are in danger of being replaced by automation and what is their educational resilience for adapting work life changes. The data of this study consist of the latest PIAAC data (The Programme for the International Assessment of Adult Competencies). Based on the research we develop a model for re-educating the people who have lost their jobs.

Keywords: Robotics · 21st century skills · Educational resilience · Work life · Adult education

1 Are Robots Replacing the Human Workforce?

Technology, automation, and robotics have already taken the place in our society. According to OECD Employment Outlook (2017) [1], the use of technology in the workplace has increased dramatically and the human workforce is replaced by the robots. The potential jobs that are at high risk of automation are transportation and storage, manufacturing, construction, administrative and support service [17].

There is a growing belief that technology innovations cause a reduction in work [2]. This means that automation and digitalisation create challenges for the work life and educational systems. It is undisputed that automation increases unemployment but it also creates new jobs and new professions [3]. Previous research opens up the effect of automation and digitalisation on employment. According to the OECD-report [1], automation will not "destroy a large number of jobs" though it is noted that low educated workers are those who will most likely suffer from the appearance of automation. "Therefore, the likely challenge for the future lies in coping with rising inequality and ensuring sufficient (re-)training especially for low qualified workers." [4].

M. Moro et al. (Eds.): Edurobotics 2018, AISC 946, pp. 197–200, 2020.
https://doi.org/10.1007/978-3-030-18141-3_16

On the other hand, there is a lot of potential in automation, digitalisation, and robotics. In the medicine, the robotics makes diagnosing faster, easier, more reliable and the operations safer [5]. In the dairy production, the health state of cows and the quality of milk can be effectively detected [6] which is safer for consumers, too. Manufacturing different items becomes easier and heavy, dangerous work executed by robots. These actions cause the reduction in work [2].

Gardner [9] compares the skills of the 20th and the 21st century and notes that the human of the 21st century has to achieve skills for using very sophisticated communication and information technologies (ICT). In general, 21st century skills include collaboration, digital literacy, citizenship, problem solving, critical thinking, creativity and productivity [10]. According to OECD [8], the 21st century skills are career and life skills, learning and innovation skills and information, media and technology skills. Strategic learning skills belong to the key skills in the 21st century due to lifelong learning, which is a necessity in working life, for individuals and organisations [11]. Further, people should possess a positive attitude and willingness to educate themselves for lifelong and width in the rapidly changing society. The idea of lifelong learning [11] also can be connected to educational resiliency, which has been used to refer to good, stable, and consistent adaption under challenging conditions [12]. Educational or academic resilience is referred to an indicator of school adjustment and a strong predictor of class participation and study motivation [13].

According to an OECD report [7], the fears of mass technological unemployment are exaggerated and they further note that there are large differences among the countries. The influence of automation is not the same everywhere and the main factor, which has the greatest equalising impact, is the skills and competencies of people. OECD [8] has published a research on the adult skills. According to report, adults who were highly proficient in the skills measured by the survey are likely to be able to make the most of the opportunities created by the technological and structural changes modern societies are going through. Still, the question is – how should we educate adults who have no previous experience of robotics.

2 Executing the Study

The aim of this research is to respond to the educational needs in the continuously and rapidly changing work-life to re-educate people to work with technology, automation, and robotics. This study is executed through mining the latest PIAAC (Programme for the International Assessment of Adult Competencies) data, which is perhaps the most far-reaching survey of adult skills. 40 countries were involved in this assessment. The measured skills are the key cognitive and workplace skills needed for individuals to participate in society and for economies to prosper. PIAAC measured adults' proficiency in literacy, numeracy, and problem solving in technology-rich environments (TRE) and gathered information and data on how adults use their skills at home, at work and in the wider community [14]. "Problem solving in technology-rich environments involves using digital technology, communication tools and networks to acquire and evaluate information, communicate with others and perform practical tasks. The first PIAAC problem solving survey will focus on the abilities to solve problems for personal, work and civic purposes by setting up appropriate goals and plans, accessing and making use of information through computers and computer networks" [15].

Altogether 5 464 adults (age 16–65 years) participated in the PIAAC 2011 research in Finland. The final participation rate was 66%. The research consisted of an interview dealing with respondents' background and assignments of cognitive skills. All the participants did the assignments independently mainly by using computers. This study focuses on the group of participants (N = 3783) who are in the labor market (full time or part time employed and unemployed). Those working at home (for example maternity leap), the students and the retired were left out of the analysis.

The target of this study is to get more information about the educational and working life backgrounds, the skills and the educational resilience of those adults, who had difficulties with problem solving in the technology-rich environment. In this study we investigate:

1. What are the skills in problem solving in technology-rich environments of the people challenged by automation?
2. What is their educational resilience for adapting work-life change?

3 Results and Discussion

- Group (N = 3783) was divided in those who were not at big risk (n = 2802) and in those who were at big risk (n = 981) with their problem solving skills in technology-rich environments (TRE). Key background characteristics of these two groups were explored and compared.
- We compared these groups and the primary results indicate that every tenth Finn who is on the labor market is at big risk and their skills in TRE are poor.
- In the group with poor skills, the share of unemployed people was 15,3%, while the respective rate was 7,1% in the group with sufficient skills.
- 76,2% of those who had poor skills were low educated (primary school or less, lower secondary, upper secondary). 60,6% of people with the good skills had tertiary education
- Individuals with poor skills were mostly working with manufacturing and construction (35,6%), trade and services (28,0%) and health and social (15,7%)
- Age distribution reveals that people with poor skills are mostly 55–65 years old (44,9%) and 35–44 years old (30,2%).
- Every tenth person (10,8%) with poor TRE skills is non-native Finnish speaker.

Based on the primary results, we suggest that the group with poor skills should be more studied to find the right forms for educating them with problem solving skills in TRE. When designing the training, we should also pay attention to non-native speakers. Training for problem solving in TRE skills can bring educational equality and give people the more competence for the world of work but also for their personal lives. There is an increasing demand for internet skills [16]. As for resilience, we need to reach people from the group of poor problem solving skills in the technology environment. Through the case study, we could get more information about the people who have defeated the challenges in this rapidly changing world of work.

Acknowledgements. This study is part of "Manufacturing 4.0 – strategies for technological, economic, educational, and social policy adoption" –project, which is funded by the Strategic Research Council (STN) of the Academy of Finland.

References

1. OECD Employment Outlook. OECD Publishing, Paris (2017). http://dx.doi.org/10.1787/empl_outlook-2017-en
2. Brynjolfsson, E., McAfee, A.: The Second Machine Age: Work, Progress, and Prosperity in a Time of Brilliant Technologies. Norton, New York (2014)
3. Autor, D.H.: Why are there still so many jobs? Hist. Future Workplace Autom. J. Econ. Perspect. **29**, 3–30 (2015)
4. Arntz, M., Gregory, T., Zierahn, U.: The risk of automation for jobs in OECD countries: a comparative analysis, In: OECD Social, Employment and Migration Working Papers, No. 189, OECD Publishing, Paris (2016). https://doi.org/10.1787/1815199X
5. Hamet, P., Tremblay, J.: Artificial intelligence in medicine, **69**(Supplement), S36–S40 (2017) https://doi.org/10.1016/j.metabol.2017.01.011
6. Butler, D., Holloway, L., Bear, C.: The impact of technological change in dairy farming: robotic milking systems and the changing role of the stockperson. J. Roy. Agric. Soc. Engl. **173**, 1–6 (2012)
7. OECD. Automation and Independent Work in a Digital Economy, Policy Brief on The Future of Work, OECD Publishing, Paris (2016)
8. OECD. Skills Matter. Further Results from the Survey of Adult Skills. OECD: Paris (2016)
9. Gardner, H.: Five minds for the future. 21st century skills. In: Bellanca, J., Brandt, R. (eds.) Rethinking How Students Learn. Bloomington. Solution Tree Press (2010)
10. Voogt, J., Roblin, N.P.: A comparative analysis of international frameworks for 21st century competences: implications for national curriculum policies. J. Curric Stud. **44**(3), 299–321 (2012). https://doi.org/10.1080/00220272.2012.668938
11. Tynjälä, P.: Perspectives into learning at the workplace. Educ. Res. Rev. **3**(2), 130–154 (2008)
12. Masten, A.S.: Resilience in Individual Development: Successful Adaptation Despite Risk and Adversity (1994)
13. Khalaf, M.A.: Validity and reliability of the academic resilience scale in Egyptian context. US-China Edu. Rev. **4**(3), 202–210 (2014)
14. OECD. Skills Outlook 2013: First Results from the Survey of Adult Skills. OECD Publishing, Paris (2013). https://doi.org/10.1787/9789264204256-en
15. OECD. Skills and Jobs in the Internet Economy, OECD Digital Economy Papers, No. 242, OECD Publishing, Paris (2014). http://dx.doi.org/10.1787/5jxvbrjm9bns-en
16. OECD, Literacy, Numeracy and Problem Solving in Technology-Rich Environments: Framework for the OECD Survey of Adult Skills. OECD Publishing, Paris (2012). https://doi.org/10.1787/9789264128859-en
17. UK Economic Outlook – Will robots steal our jobs? The potential impact of automation on the UK and other major economies (2017)

Teaching Object-Oriented Programming in Secondary Schools Using Swarm Robotics

James Stovold[1(✉)] and Stewart Powell[1,2]

[1] Department of Computer Science, Swansea University, Swansea, UK
{j.h.stovold,s.w.powell}@swansea.ac.uk
[2] Technocamps, Swansea University, Swansea, UK

Abstract. The recent inclusion of computer science in the British secondary school education system has resulted in existing teaching staff who are not able to effectively deliver the curriculum. Existing environments—such as Greenfoot—help substantially but anecdotal evidence suggests that many pupils still struggle with some aspects of the computer science curriculum. This paper presents a workshop for teaching pupils about method calls in object-oriented programming, using swarm robotics and the firefly synchronisation algorithm as inspiration.

Keywords: Teaching object-oriented programming · Swarm robotics · Greenfoot · Firefly synchronisation algorithm

1 Motivation

The redevelopment of national curricula in the UK has resulted in a resurgence of computer science within primary and secondary schools [2]. The devolved education systems across the UK mean that the implementation of computer science is not consistent. Following an independent review of the Welsh national curriculum [5], a new national school curriculum is in development for Wales [13] with computer science as a key focus. Currently there is a particular struggle for secondary schools in resource-constrained communities—such as the many isolated towns and villages in Wales—to teach computer science.

Technocamps [4] is a schools outreach programme that started in 2003 in the Computer Science department at Swansea University. It specifically focusses on how to overcome the issues around teaching computer science within schools, by providing in-school workshops to motivate and inspire pupils, as well as providing many opportunities for continuing professional development for teachers responsible for teaching computer science.

Technocamps facilitators work closely with teachers during these workshops and through this process have received anecdotal evidence that the teachers are struggling to teach object-oriented programming. The WJEC Computer Science GCSE [14] specifies that pupils use the Greenfoot programming environment [7],

M. Moro et al. (Eds.): Edurobotics 2018, AISC 946, pp. 201–204, 2020.
https://doi.org/10.1007/978-3-030-18141-3_17

but many teachers still report that pupils struggle to grasp certain concepts—most notably, method calls.

Through discussions with pupils, the problem appears to be a reluctance to think in terms of the object—instead 'playing god' and thinking in terms of a global observer. This inspired us to think of a way to switch this perspective, so we could demonstrate method calls without discussing objects and classes directly. The field of swarm robotics is ideal for this task, allowing us to use tangible objects (robots) which rely heavily on method calls (inter-robot communication) to operate.

This paper presents details of a workshop we use to teach pupils (aged 14 to 16 years old) about method calls through swarm robotics. Section 2 discusses swarm robotics along with providing details of the programming environment Greenfoot [7] that is required by the GCSE specification. Section 3 presents the firefly algorithm and details of the workshop we run with the pupils. Finally, Sect. 4 discusses the impact of the workshop on the pupils' learning, and presents our intentions for future work in this area.

2 Swarm Robotics and Greenfoot

Swarm robotics are a form of multi-robot system that make use of emergent behaviour to achieve their goal [1,10]. The robots are typically simple, comparatively cheap, with limited sensing ability. The algorithms that govern their behaviour often take inspiration from social insects [6,8] and rely heavily on the interaction between robots to produce the desired behaviour.

One such swarm behaviour is the *firefly synchronisation algorithm* [12] which uses basic behaviour (robots flashing) to communicate and to synchronise the robots.

The reliance on interaction by swarm robots make them the ideal candidate for describing method calls in object-oriented programming. The aim is to get pupils to think in terms of the robot, rather than an observer—how does a robot request something of another robot? How does a robot know when another robot has flashed?

Greenfoot [7] provides an environment and an 'actor' class that the pupils program to interact with the environment. Pupils have the option to include multiple actors in the same environment, and have them interact with each other through method calls, but—as described above—the majority of pupils appear to struggle with this concept.

To address this knowledge gap, we developed a new layer that sits atop the Greenfoot 'actor' class. This layer implements basic robot functionality such as requesting all other robots that are within a given radius (i.e. the `neighbours()` function). By providing this simple functionality we aim to lower the barrier to entry for pupils, making the Greenfoot exercises more accessible while linking them to real-world applications through the use of robots. This layer is provided free-of-charge through our github account (http://github.com/jstovold/greenfoot).

3 Workshop

The workshop we run with pupils is centred around the *firefly synchronisation algorithm* [12]. This is a simple synchronisation algorithm that can be implemented in groups of robots, groups of pupils, or groups of Greenfoot 'actors', making it ideal for our purposes.

Fireflies exhibit a particularly interesting behaviour: individual male fireflies will flash on a periodic basis to attract a female. When collected together into a swarm, the flashing of the fireflies will synchronise such that all flashes occur simultaneously [3]. This increases the overall brightness of the flashing, meaning the males are more likely to attract females, benefitting all those involved.

This behaviour has since been adapted into an algorithm for synchronising distributed components of artificial systems (such as a swarm of robots) [9,12], and has even been used to form identities in artificial swarms [11].

The firefly synchronisation algorithm [12] is remarkably simple. Each robot maintains an internal counter and increments that counter on a regular basis. Once the counter reaches a predefined threshold, the robot flashes and resets its counter to zero. If a robot sees another robot flash when it is not flashing, it will increase its counter by two instead of one for that step. Over time, this is sufficient to synchronise the flashing of all robots that can see each other.

During our workshop with the pupils, we describe the ideas behind swarm robots, social insects, and show the behaviour of fireflies in the wild. Next, we ask the pupils to move into small groups (max. 4 pupils) and discuss how they would implement this synchronisation for robots. In every session we have run, the pupils come up with similar ideas: watch for other robots, when they flash, you flash. While this would not result in synchronised flashing, it is an important step for the pupils as they are now actively thinking from the perspective of a robot, rather than the observer.

After 10 min of discussion, we describe the firefly algorithm. Next, we give the basic rules of the algorithm to the pupils, and get them to act it out:

1. Pick a number (0–15)
2. Start counting (on each beat)
3. When you hit the maximum (15), flash (raise your hand)
4. Start again from 0
5. If you see somebody else flash (raise their hand), skip a number

We take the pupils out into an empty area and divide them into groups of around 10, and assign a facilitator to provide the 'beat' for them—in this case, regular periodic clapping.

Over time, the pupils start to raise their hand at the same time—they have synchronised. Once they have seen that such a simple algorithm can work, we demonstrate the robots running the same algorithm (either live or using a recording), and discuss why synchronisation between robots might be useful. Finally, the pupils are asked to go to their computers and implement this algorithm using Greenfoot and our new software layer. By this point, they will have seen the underlying natural phenomena, have acted out the algorithm, and seen how this relates to real robots and the importance of synchronising their behaviour.

4 Discussion

Through informal discussions with teachers and pupils after running the workshop, it appears that pupils take to this idea without realising that they are implementing method calls. While this work is still in its early stages, these positive interactions give us hope that this will act as a stepping-stone for pupils and teachers alike to understand method calls, and be able to teach this effectively moving forwards.

Our future work includes developing a survey to quantify the efficacy of the new teaching method as well as evaluate the positive impact possible and teachers and their pupils alike.

References

1. Beni, G., Wang, J.: Swarm intelligence in cellular robotic systems. In: Robots and Biological Systems: Towards a New Bionics? NATO ASI Series, vol. 102, pp. 703–712. Springer, Heidelberg (1993)
2. Brown, N.C.C., Sentance, S., Crick, T., Humphreys, S.: Restart: the resurgence of computer science in UK schools. Trans. Comput. Educ. **14**(2), 9:1–9:22 (2014)
3. Buck, J.B.: Synchronous rhythmic flashing of fireflies. Q. Rev. Biol. **13**(3), 301–314 (1938)
4. Crick, T., Moller, F.: A national engagement model for developing computer science education in wales. In: Proceedings of ISSEP 2016 (2016)
5. Donaldson, G.: Successful Futures: Independent Review of Curriculum and Assessment Arrangements in Wales (2015)
6. Dorigo, M., Maniezzo, V., Colorni, A.: Ant system: optimization by a colony of cooperating agents. IEEE Trans. Syst. Man Cybern. Part B Cybern. **26**(1), 29–41 (1996)
7. Kölling, M.: The greenfoot programming environment. Trans. Comput. Educ. **10**(4), 14:1–14:21 (2010)
8. Millonas, M.M.: Swarms, phase transitions, and collective intelligence. In: Langton, C.G. (ed.) Artificial Life III, pp. 417–445. Addison-Wesley, Boston (1993)
9. Mirollo, R., Strogatz, S.: Synchronization of pulse-coupled biological oscillators. SIAM J. Appl. Math. **50**(6), 1645–1662 (1990)
10. Şahin, E.: Swarm robotics: from sources of inspiration to domains of application. In: Şahin, E., Spears, W. (eds.) Swarm Robotics. Lecture Notes in Computer Science, vol. 3342, pp. 10–20. Springer, Heidelberg (2005)
11. Stovold, J., O'Keefe, S., Timmis, J.: Preserving swarm identity over time. In: Sayama, H., Rieffel, J., Risi, S., Doursat, R., Lipson, H. (eds.) Proceedings of the 14th International Conference on the Synthesis and Simulation of Living Systems (ALIFE 2014), pp. 728–735 (2014)
12. Tyrrell, A., Auer, G., Bettstetter, C.: Firefly synchronization in ad-hoc networks. In: Proceedings of the MiNEMA Workshop (2006)
13. Welsh Government: Education in Wales: Our national mission. Action plan 2017–21 (2017)
14. Welsh Joint Education Committee: WJEC GCSE in Computer Science-Specification 2017–2019 (2016)

Promoting Creative Thinking and Problem Solving Through Robotic Summer Camp

Ana Sovic Krzic, Liljana Puskar, and Tomislav Jagust(✉)

Faculty of Electrical Engineering and Computing, University of Zagreb, Unska 3, 10000 Zagreb, Croatia
{ana.sovic.krzic, liljana.pushkar, tomislav.jagust}@fer.hr

Abstract. In this paper we present our experience in organizing a robotic summer camp for pupils from 9 to 15 years old. During the camp, participants had to solve a number of robot construction and programming tasks, designed in a way that encouraged them to develop their creative thinking and problem solving skills, like creating a robot from a pre-defined set of LEGO pieces or building a faster paper plane launcher by using a gear transmission system. The results of the conducted survey showed that, even though pupils presented a solid level of creativity and problem solving skills, most of them found the lessons that fostered creative thinking and problem solving skills to be quite challenging.

Keywords: Educational robotics · STEAM · Problem solving · Creative thinking

1 Introduction

Creative thinking and problem solving are two of the most important 21st century skills, necessary to successfully face challenges and needs of the modern society, as well as future job market [1, 2]. It is believed that educational robotics could be a successful tool for development of these skills [3, 4]. On the other hand, many school programs and curricula are still not aligned with the needs of the digital natives they have to educate. This gap can be filled by out-of-school non-formal or informal educational activities like different clubs or summer camps, where pupils can freely and relaxed, without the burden of formal learning objective or a graded test, develop their 21st century skills [5]. In this paper we present our experience in organizing one of these summer camps, where 22 pupils, surrounded by colleagues with similar age and interests solved a number of robotics challenges and practiced different skills, from creative thinking and problem solving to group work, collaboration and knowledge transfer.

Using educational robots in and out of class has been on a rise. Adams et al. [6] received a positive feedback using Lego Mindstorm NXT in generating and solving different problems. Theodosi and Deli [7] combined art and robotics in a student Robotics Club - on the basis of a poem, 5th and 6th grade students built a robotic horse using Lego Mindstorm NXT. LEGO Mindstorms were also used to teach programming [8], boosting interest, engagement, creativity and willingness of students to learn

M. Moro et al. (Eds.): Edurobotics 2018, AISC 946, pp. 205–208, 2020.
https://doi.org/10.1007/978-3-030-18141-3_18

programming. After facing problems in attracting students for engineering, Ayar et al. [9] developed a robotics summer camp to bring STEM closer to high school students. Our previous work [10] shows that gifted children can use LEGO Mindstorms as a learning tool, and while learning they display an aptness for creativity in building and programming robots.

2 Summer Robotic Camp

Croatian robotic association has organized Summer robotics camp "Petica" since 2016. The camp, intended for pupils between 9 and 15 years, usually lasts for 5 days and is organized in a small countryside farm hostel. This year, 22 students (7 girls and 15 boys) participated in the camp, and were divided in two groups: beginners (5 girls and 6 boys) and experienced (2 girls and 9 boys). Students took Lego robotic lessons during morning (each activity lasted for 2 h), and different physical activities (like countryside treasure hunt or Backyard Olympics) during afternoons and evenings. All activities gradually introduce new terms and concepts of mechanics and programing. Tasks logically follow up each other, strengthen the knowledge of the past activities and add a new value.

Since the beginners did not have any significant experience in robotics and did not work with LEGO robots before, their starting tasks were rather simple: construct the robot using the given instructions and learn how to program it. Some of the challenges were: (1) make robot to spin around a bottle, (2) drive slalom between bottles, (3) move the LEGO figure/model to the designated position, (4) using ultrasonic sensor find the way out of the maze, and (5) recognize the given color. All those activities strengthen the problem-solving skills, since the students knew the anticipated behavior of the robot and had to come up with the program that enables it. Later on, the tasks became more open-ended, fostering creative solutions, like (6) make an alarm using ultrasonic sensor, (7) use gears to create mechanism for displaying cartoons, (8) make a cartoon, and (9) create a LEGO doodler.

The experience, who already participated in robotics workshops and competitions, started with more complicated and creative tasks, without precise instructions: (1) build a robot with maximum of 24 LEGO pieces (Fig. 1), (2) make a sweeper that cleans the floor of LEGO bricks (Fig. 2) [11], and (3) upgrade the sweeper with the IR remote control. In the second part of the camp, experienced group focused on learning more advanced programing methods using Space challenge [12] as a background story. Among others, students solved missions: (4) free the MSL robot, (5) activate communication, (6) launch the rocket, and (7) use the solar cells. On the last day, students learned how to (8) follow a black line and go around the obstacles and (9) create the paper plane launcher by using previous knowledge about gears (Fig. 3).

On the last day of the Camp, students were given a short survey (Fig. 4). Here, due to the page limit, we present only students' evaluations of the activities. In survey, 1 stands for "to complicated" and 5 stands for "interesting and fun". The most difficult task for the "beginner group" was "gears", where students learnt basics of gears and had to use the knowledge to develop a mechanism for displaying cartoons, without instructions. The most fun tasks were "animation" where robot showed students' pictures and "doodler" where robots plotted colorful pictures and students had to make simple programs or modification in programs/robots.

The best rated activities in experienced group are "black line follower", "free the MSL robot" and "activate communication". In those activities students built the robot using the instructions but had to design their own program. Since almost all of the students solved these tasks, they showed their well-developed problem-solving skill. The worst rated activities are "24-piece robot", "paper plane launcher" and "sweeper". In those activities students did not have building instructions. They had to use their imagination and creativity to solve the task. Although "24-piece robot" and "sweeper" were solved by all students, many of them showed some level of frustration because they did not have building instructions. In such cases, other students helped with brainstorming and demonstration of their solutions. The "paper plane launcher" was the most complicated task, only one student managed to make a fully functional plane launcher.

Fig. 1. Robots with less than 24 pieces.

Fig. 2. Robot sweeper.

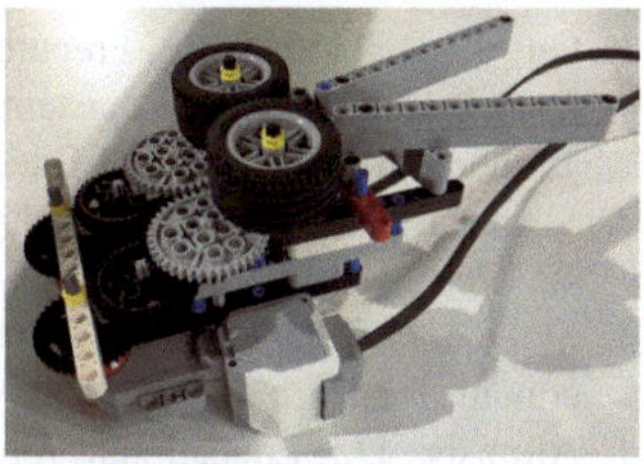

Fig. 3. Paper plane launcher.

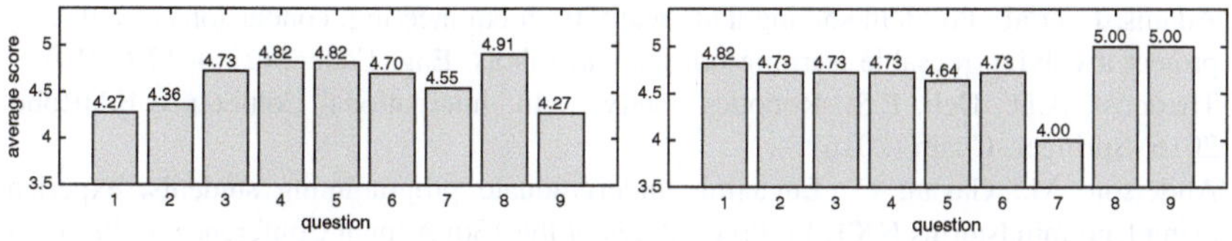

Fig. 4. Average scores in students' survey. Left - experienced group: (1) 24-piece robot; (2) sweeper; (3) sweeper using IR remote control; (4) free the MSL robot; (5) activate communication; (6) rocket launcher; (7) solar cells; (8) black line follower; (9) paper plane launcher. Right - beginners group: (1) spinning around bottle; (2) slalom; (3) animal rescue; (4) maze; (5) color recognition; (6) alarm; (7) gears; (8) animation; (9) doodler.

3 Conclusion and Future Work

In this paper we briefly described our experience and some of the activities that were the part of 5-day robotics summer camp for primary school pupils in Croatia. Participants engaged in a number of different activities that taught them basics of robotics and programming, but also fostered their creative thinking, problem solving, collaboration and knowledge transfer skills. Based on the survey results and behavior of students, both groups, beginners and experienced, showed that they have improved their problem solving skills in robot programming, but still had some issues with building different robots. Even though pupils seemed engaged and interested, they found the lessons that fostered creative thinking and problem solving skills to be quite challenging and different from the lessons they were accustomed at in school. In the future, we plan to further improve the existing activities to make them more acceptable to students, at the same time developing their 21st century skills.

Acknowledgment. This work has been supported in part by Croatian Science Foundation under the project UIP-2017-05-5917, University of Zagreb and Croatian robotic association.

References

1. Gu, J., Belland, B.R.: Preparing students with 21st century skills: integrating scientific knowledge, skills, and epistemic beliefs in middle school science curricula. In: Emerging Technologies for STEAM Education. Springer, Switzerland, pp. 39–60 (2015)
2. Shaheen, R.: Creativity and education. Creative Educ. **1**, 166–169 (2010). https://doi.org/10.4236/ce.2010.13026
3. McDonald, S., Howell, J.: Watching, creating and achieving: creative technologies as a conduit for learning in the early years. Br. J. Educ. Technol. **43**(4), 641–651 (2012)
4. Alimisis, D.: Educational robotics: open questions and new challenges. Themes Sci. Technol. Educ. **6**(1), 63–71 (2013)
5. Cavas, B., et al.: The effects of robotics club on the students' performance on science process & scientific creativity skills and perceptions on robots, human and society. In: Proceedings of 3rd International Workshop Teaching Robotics, Teaching with Robotics Integrating Robotics in School Curriculum (2012)
6. Adams, J., et al.: Problem solving and creativity in engineering: conclusions of a three year project involving reusable learning objects and robots. Eng. Educ. **5**(2), 4–17 (2010)
7. Theodosi, A.D., Deli, E.S.: Robotics poetry…. In: International Conference EduRobotics 2016. Springer, Cham (2016)
8. Anderson, M., Gavan, C.: Engaging undergraduate programming students: experiences using lego mindstorms NXT. In: Proceedings of the 13th Annual Conference on Information Technology Education. ACM (2012)
9. Ayar, M., et al. A robotics summer camp for high school students: pipelines activities promoting careers in engineering fields. In: 2013 ASEE Annual Conference. American Society for Engineering Education (2013)
10. Jagust, T., Cvetkovic Lay, J., Krzic, A.S., Sersic, D.: Using robotics to foster creativity in early gifted education. In: International Conference on Robotics and Education RiE 2017. Springer, Cham (2017)
11. Center for Engineering and Education Outreach, Tufts University. https://mindstorms.dreschallenges.com. Accessed 28 Jul 2018
12. The Lego group. https://education.lego.com/en-us/support/mindstorms-ev3/space-challenge. Accessed 28 Jul 2018

From Concept Inventories to Robotic Competitions

Reinhard Gerndt[1(✉)] and Jens Lüssem[2]

[1] Ostfalia University of Applied Sciences, Wolfenbüttel, Germany
r.gerndt@ostfalia.de
[2] Kiel University of Applied Sciences, Kiel, Germany
jens.luessem@fh-kiel.de

Abstract. Robotics competitions play a crucial role in current tertiary education and in research. In education, competitions typically are held to activate passive knowledge. Research competitions serve as a testbeds and as benchmarking for new ideas. With this paper we show how we systematically derived an Open Humanoid Competition rule set from a concept inventory, that covers the central concepts of a field as agreed on by a the respective scientific community.

Keywords: Concept inventory · Robotics competitions · Open humanoid competition

1 Introduction

Teaching robotics is a highly trans-disciplinary field of education. Robotics competitions create a learning environment independent of a class room situation and are thus a very suitable measure for efficient teaching and acquisition of in-deep knowledge. However, the design of robotics competitions often appears to be random, inspired by a narrow research question or scientific hype of the time. Thus the competitions often are centered on a very narrow task, specific robots or robot kits, as can be inferred from the participants. On the other side, with concept inventories (CIs) [1–3], there is a solid starting point for the systematic definition and refinement of competitions. The proposed competition will thus be more independent of background and experience of participants and should yield an open test bed for a multitude of humanoid robotics and production technologies and a vibrant exchange in the community.

2 Concept Inventories

CIs have been introduced as means of assessing learning outcomes and controlling teaching. They list the core concepts of the respective scientific field and include a test to elicit the knowledge about them. By comparing test results before and after a lecture a learning gain can be calculated to assess the quality of the

M. Moro et al. (Eds.): Edurobotics 2018, AISC 946, pp. 209–213, 2020.
https://doi.org/10.1007/978-3-030-18141-3_19

Table 1. Robotics Concept Inventory.

No.	Category	Concept(s)
1	Math and numerical methods	Reference systems, approximation, signal processing
2	Mechanics	Basic mechanical principles, basic mechanical building blocks
3	Kinematics	Basic kinematic building blocks, forward kinematic, inverse kinematic
4	Stability	Center of mass and support polygon
5	Dynamics	Rigid body dynamics, rotational movement
6	Control Theory	Feedback, limits (quality/accuracy) of control systems
7	Sensing	Time in sensing, quantification and measurement accuracy
8	Perception	Object properties, complex environments
9	Behavior	Reactive behavior, sequential behavior
10	Decision-making and planning	Metrics for decision-making, strategies, complex plans, conflicts
11	Uncertainty	Probabilities, Reality/model relation

respective teaching method. The specific initial robotics CI [3] consists of 11 categories and 25 concepts or sub-concepts as sketched in Table 1.

The challenge of systematically defining competitions is allowing for different levels of experience of the envisioned participants and at the same time allowing a benchmarking among participants and over time. Participants may e.g. stem from mechanical engineering or computer science and may be undergraduate or graduate students. The CIs provide a valuable base to derive the requirements, as they do not refer to any specific skill. For example, the first robotics concept of Table 1 is related to the coordinate transformation between different frames of reference, possibly a global (inertial) and a local (mobile) one. The transformation concept has different levels of complexity. In mechanics it could be a displacement along a single or multiple axes, a rotation around a single or multiple axes or a combination thereof. Image acquisition and processing additionally requires scaling and perspective. The concept-based competition design thus assures that different levels of complexity, interests and backgrounds are considered.

3 The Open Humanoid Competition

The Open Humanoid Competition is based on the multi-dimensional scenario of bipedal robots playing soccer and on the robotics concepts as listed in the robotics CI. By verifying the requirements and constraints as requested by the standard soccer rules against the robotics CI, the number of constraints and requirements could be minimized, leaving maximum room for participants creativity without sacrificing the ability to benchmark the research outcomes.

Robotic soccer requires locomotion and ball handling. It may also require articulating the arms, as for getting up from a fall or for a goal keeper. Both aspects are reflected by kinematics category of the CI. However, only one aspect actually is required. Therefore, rules where defined, such that articulation of the arms is not required by allowing to design the robots such that locomotion is sufficiently robust, e.g. by sufficiently large feet, low center of mass or sophisticated control algorithms. In contrast RoboCup Humanoid League rules require a human-like center of mass (COM) that significantly affects the robustness of walking. This way, a humanoid robot can be implemented with as little as 4 motors, Fig. 1 (left). Similarly the many ball handling options, like pushing, dribbling or kicking do not differ significantly with respect to the CI concepts. Therefore, rules request no specific capability, like kicking, as it is the case in the regular RoboCup rules. In order to have the same task for all robots, which will also ease the benchmarking, a cardboard goal keeper is foreseen, Fig. 1 (center).

Fig. 1. Open Humanoid Soccer Robot (left), cardboard goal keeper (center), local map (right).

To cover perception-related concepts, we propose to code ball and goalkeepers with dedicated colors, Fig. 1 (right), which allows to identify the main objects in the soccer scenario with basic algorithms, e.g. color blob detection or dedicated devices like smart cameras, e.g. CMUCam (PixyCam). Additionally identifying the robot players causes additional effort but may be advantageous.

There are multiple complexity levels in localization. An ego-centric model is a very basic approach of relative localization, Fig. 1 (right), which still covers the respective concept categories like Perception, behavior and decision-making. More mature teams may introduce advanced approaches, e.g. based on derivatives.

The derived minimal set of rules fits on a single sheet of paper (Fig. 2). The first part defines the field, the goals and the ball as explained above. A game is played by robot teams of 3–6 robots from different participants. The range of number of robots per team allows for easy adjustment to the actual number of participants, such that competition with multiple games can be held with as little as 6 participants. Constraints on the robots, other than bipedal locomotion, are reduced to an minimum. For practical and safety reasons weight and size are limited. There is no restrictions on the COM height and foot size. Robots may be started remotely at any position in the field or manually at the touch line, such that human robot handlers do not need to enter the field during start or restart. The game consists of two 10-minutes halves with a 5 min break. Start

Open Humanoid Robot Soccer Competition (OHRSC)
The games in the (OHRSC) are played as drop-in games for bipedal robots, with a minimal rule set. Every participating individual or group contributes a single robotic player to a robot team. Robots are scored according to goals scored directly (2 points) and assists by being in the field during a goal for the own team (1 point). The Robot with the highest score wins the competition.

1. **Field of Play**
   The OHRSC is played on an field with a maximum extension of 18 * 6 meters. The size of the field may be adjusted but shall not change during the competition. Cyan, respectively magenta cardboard figures represent the goal keepers. They are suspended from the respective goals cross bars, reaching the floor. Both figures have the same form. They are positioned by the referee, who may change their position during any stoppage of the game.
2. **The Ball**
   The ball is a FIFA size 1 ball with orange colored surface.
3. **The Number of Players**
   The robot teams consist of 3 – 6 robots. Players are chosen randomly for every single game.
4. **The Players**
   The robot players are bipedal autonomous robots that walk on two legs and drive a ball by pushing or kicking. The maximum size of the robots is 1 meter, the maximum weight is 10 kg. Robots must bring red and blue team markers and must not use blue or red on any part that is not interchangeable. Robots may be started manually at the touch line of their team's own half. Robots that are started remotely by a game controller may be positioned anywhere in their own half
   After positioning and possibly starting the robots, human team members must not interfere with the game play, except if explicitly called for by the referee. Robots need to be equipped with a handle to be safely picked up by a referee and an emergency switch-off that is clearly marked and reachable. Pick-up of robots by robot handler and re-entry only is allowed during stoppage of the game.
5. **The Referee**
   Each match is controlled by a team of referees with one head referee, who has full authority to enforce the Laws of the Game.
6. **The Duration of the Match**
   A match is played in two half times of 10 minutes each with a break of 5 minutes.
7. **Start and Restart of Play**
   The play is started and restarted by a drop ball at the kick-off point at the center. The ball is in play for both teams immediately. Goals may be scored immediately.
   A game is restarted for:
   - second half time,
   - after a goal was scored,
   - after more than half of the robots of a team became inactive or removed after a (re-)start,
   - stuck game.

   For (re-)start of game robots may be placed and possibly manually started by a robot handler.
8. **The Ball In and Out of Play**
   If the ball leaves the field of play, it is returned to field by the referee or an assistant referee, one meter into the field. There is no stoppage of the game.
9. **The Method of Scoring**
   A goal is scored if the ball completely crossed the goal line between the two goal posts.
10. **Fouls and Misconducts**
    The following cases are considered as fouls, resulting in a removal penalty until next restart:
    - any action posing the potential to significantly damaging a robot, including itself,
    - leaving the playing field (field of play and surrounding green surface),
    - any other significant offence, e.g. obstructing the game, as considered by the referee.

    Robots potentially endangering humans by whatever activity are excluded from the game (red card).

Fig. 2. Open Humanoid Competition rule set fits on one page.

and restart of the game is done by a drop ball at the center point. The game continues untiltime is up, a goal is scored or until there is no further progress in the game (stuck game) for whatever reason. Then there is a restart. Disallowing to pick up fallen robots, other than during a stoppage of the game, is considered a suitable rule to foster the development of robust robots. If the robot is in danger of damaging itself or others, it will be removed until next restart. Scoring is primarily based on scoring goals and supporting other robots for scoring goals. As a default rule, significant authority is given to the referee.

The Open Humanoid Competitions shall be co-located with a regular competition, e.g. the RoboCup Humanoid League to use existing infrastructure.

4 Conclusions

We demonstrated how the Open Humanoid Competition as an interdisciplinary competition for soccer robots was derived systematically from the robotics concept inventory. It provides an educational scenario, motivating new students through manageable, entry-level challenges and entertaining competitions and provides a testbed for advanced students with more sophisticated robot designs.

References

1. Wage, K.E., et al.: The signal and systems concept inventory. IEEE Trans. Educ. **48**(3), 448–461 (2005)
2. Lindell, R.S., et al.: Are they all created equal? A comparison of different concept inventory development methodologies. PERC Proc. **883**, 14–17 (2006)
3. Gerndt, R., Lüssem, J.: The robotics concept inventory. In: International Robotics in Education Conference (RiE) (2018)

Teaching Physics Concepts Using Educational Robotics

Domenico Guastella[1,2(✉)] and Antonella D'Amico[1,2]

[1] Dipartimento di Scienze Psicologiche, Pedagogiche e della Formazione dell'Università degli Studi di Palermo, Palermo, Italy
{domenico.guastella,antonella.damico}@unipa.it
[2] MetaINTELLIGENZE ONLUS - Centro Studi Internazionale, Palermo, Italy

Abstract. This paper reports a study included in the national project PROMISE funded by the Italian Ministry of Education, University and Research. It examines the application of an educational robotics laboratory with secondary school students for learning concept about physics. In the paper we describe the didactic strategies that have been implemented during the activities and we give some qualitative outcomes related to effectiveness of the laboratory. Indeed we observed interesting results in students both in term of learning and of interest and participation. The importance of teacher involvement in this type of intervention will also be discussed.

Keywords: Educational robotics · Physics teaching/learning · Cooperative learning

1 Robots as Educational Tool

In recent years, Educational Robotics (ER) has been used for developing many competencies: several studies have shown that educational robotics is positively correlated with cooperative learning and the sense of individual and collective self-efficacy [1]. Other studies showed that the students who used the robots achieve better results in learning concepts such as space and distance [2], geometry [3], spatio-temporal relations [4] and Genetics [5]. Caci, D'Amico and Chiazzese [6] demonstrated also that robotics labs may improve specific cognitive skills such as visual-spatial working memory. Moreover, all these studies evidenced that students are very collaborative and motivated in carrying out the robotic activities.

2 The PROMISE Project

Starting from these results, we performed a study aimed at use ER as a teaching/learning support tool for the didactic of Physics.

The study was carried out inside the national project "PROMISE" (Progetto RObotica Educativa: Musei, Imprese e ScuolE). PROMISE has been funded by MIUR (Italian Ministry of Education, University and Research) and realized thanks to the

M. Moro et al. (Eds.): Edurobotics 2018, AISC 946, pp. 214–218, 2020.
https://doi.org/10.1007/978-3-030-18141-3_20

collaboration of many public and private entities such as: Scuola Superiore Sant'Anna (lead institution), University of Milan-Bicocca, CNR (Institute of Educational Technologies of Palermo), University of Siena, University of Palermo, MetaIntelligenze Onlus, PalermoScienza, National Network of Schools for Robocup Junior Italy. The project aimed to create a national network of educational robotics involving schools, museums, and companies and, in particular, to train 36 teachers and 24 experts in the use of educational robotics in learning contexts. Both teachers and experts attended a training course and also performed a 36-h internship for applying the skills acquired at school or in museums.

3 Teaching Physics Using Educational Robotics

Our study concerned the use of robotics for teaching/learning concepts of physics related to energy and motion. A class of 21 (mean age 16 years), attending the 3rd year of a scientific secondary school, participated to the study. The activities were conducted by their math and science teacher, who participated to the PROMISE Project. An experimenter from our team followed all the activities as observer.

Five LEGO Mindstorm EV3 kits and 5 PCs were used in the school's computer lab. The 21 students were divided in 5 groups, so that each group had the opportunity to use one robotic construction kit. The ER activities were carried out in 6 weekly meetings lasting two-hour each.

During the meetings the students studied some topics about physics using ER. They realized several experiments that included the application of formulas on motion and energy for programming the movements of robots for solving some problems, as described below. At the end of each meeting, the students showed the programs to the other classmates, exposed the main problems they had encountered and illustrated the solutions they had found to solve them.

3.1 First Meeting

During the first meeting the students took a look at the main programming blocks related to the movement and use of the sensors. After a brief introduction, they were left free to perform two simple robot movement programs by using sensors to avoid obstacles or to stop when a specific condition occurred.

3.2 Second Meeting

The second meeting focused mainly on the use of "flow" programming blocks (loop, switch, wait …) and "data operation" (variable definition, mathematical operations and arrays) necessary to apply the motion formulas in subsequent meetings. The students followed a tutorial and created a program that used these functions.

3.3 Third Meeting

During the third meeting, after the study of formulas to calculate the motion and the relative inverse formulas, each group had to solve a problem about physics by using robots. The first step was to calculate the robot speed, that is not defined in EV3 construction kits. In order to achieve the focus, they used a meter for calculating the length of a path and a stopwatch for measuring the time that the robot needed to follow it. Once students owned information about space and time, they could derive the robot speed. This information might then be used to predict the travel time of the robot along a straight path or, applying the inverse formulas of the uniform rectilinear motion, for measuring the length of different surfaces.

3.4 Fourth Meeting

During the fourth meeting the students were asked to implement a program that calculated the speed and acceleration of the robot through the radius of the wheels and the number of rotations of the engine. By using the "data operation" programming blocks, the robot can be programmed to perform calculations, by using variables (in this case the time and the number of rotations) and by displaying the results on the computer programming interface or in the small display that is embodied in the robot. Every time the engine powers and the distance traveled are modified, the program calculates and displays the new speed and acceleration.

3.5 Fifth Meeting

During the fifth meeting the students were asked to calculate the speed of a rotating rod attached to a robot wheel (whose radius was known). In order to make the task more complex, the students were also asked to calculate the acceleration of the rod. This was made to activate their critical reasoning as students had to realize that the calculation could not be done with the available data.

3.6 Sixth Meeting

During the last meeting the students created a program at their choice by using the formulas or programming blocks used in the previous meetings and choosing to work on the concepts of physics or programming.

They created interesting programs: Group 1 created a robot that calculated its length, if dragged on a surface; Group 2 created a robot able to carry out a path without impacting walls or obstacles; Group 3 made a variant of the robot that calculates the speed without using the stopwatch, but setting a timer so that the robot would stop after a predetermined period of time; Group 4 made a variant of the robot that calculates the distance, that was able to display the results of the calculations on the embodied display and keep in memory the previous measurement; Group 5 created a robot capable of recognizing the color of a spherical object and of transporting it to a particular position.

4 Discussion

During our experience, that is only a limited part of the project PROMISE, we could ascertain that ER may significantly contribute to creating a stimulating learning environment for both students and teachers.

Concernig to the students, as other studies already demonstrated, there is a particular interest towards the robots as well as towards all that is digital and technological. The observation of the dynamics that have been established during the workshop path demonstrated a strong participation by the whole class group with evident stimulation of the ability to work in group, an intense collaboration and a continuous comparison among pairs. Finally, observing the results of the last meeting, in which the students were free to create a program, we can see how the skills and abilities that were stimulated during this type of activity were extremely various. In fact, we can see that some groups have focused more on the implementation of programs that solve problems in physics, others instead have worked more on problem-solving related to the programming of the robot (i.e. to walk a track avoiding obstacles or recognizing colors), others have also focused on "engineering" aspects, such as adapting the robot body according to the task to be performed. Another significant outcome that resulted by the final interviews with the teacher is the perception of an increasing in the academic performance of the students that were usually less motivated to learning.

In general, also the teacher was very satisfied about the activity, and she wondered about the improvement in learning of her students.

More in general, we can affirm that all teachers involved in the PROMISE project have been showing a strong interest and motivation about ER. As demonstrated by the talks with the teacher during and after the internship, this type of activity improves the feeling of effectiveness of their teaching method, making teachers more motivated and improving the perception of their work status. Indeed, they not only consider robotics an excellent tool for support teaching but they also see it as a big change compared to the monotony of traditional teaching. The greater involvement of students in the topics dealt using the robots, makes the teacher's work more rewarding. It makes easier keeping the class's attention for long-term.

In the light of these results we can affirm that educational robotics is not only a facilitator for students but also for teachers. The introduction of these tools to support teaching seems to have positive effects also on the school climate and the performance of teachers within their working context.

Our aim, in the following studies and activities of the Promise Project is to sustain all these claims reporting also quantitative data, that are currently under collection.

References

1. Kanda, T., Shimada, M., Koizumi, S.: Children learning with a social robot. In: 2012 7th ACM/IEEE International Conference on Human-Robot Interaction (HRI), pp. 351–358. IEEE (2012)
2. Mitnik, R., Nussbaum, M., Soto, A.: An autonomous educational mobile robot mediator. Auton. Robots **25**(4), 367–382 (2008)

3. Grimaldi, R., Grimaldi, B.S., Marcianò, G., Siega, S., Palmieri, S.: Robotica educativa e potenziamento delle abilità visuo-spaziali. In: Didamatica 2012. Informatica per la didattica, vol. 1, pp. 1–10. AICA-Università di Bari (2012)
4. Julià, C., Antolí, J.Ò.: Spatial ability learning through educational robotics. Int. J. Technol. Des. Educ. **26**(2), 185–203 (2016)
5. Baker, M., De Vries, E., Lund, K., Quignard, M.: Computer-mediated epistemic interactions for co-constructing scientific notions: lessons learned from a five-year research programme. In: Proceedings of EuroCSCL, pp. 89–96. (2001)
6. Caci, B., D'Amico, A., Chiazzese, G.: Robotics and virtual worlds: an experiential learning lab. In: Biologically Inspired Cognitive Architectures 2012, pp. 83–87. Springer, Heidelberg (2013)

Author Index

M. Moro et al. (Eds.): Edurobotics 2018, AISC 946, pp. 219–220, 2020.
https://doi.org/10.1007/978-3-030-18141-3

MIX
Papier aus verantwortungsvollen Quellen
Paper from responsible sources
FSC® C105338

If you have any concerns about our products,
you can contact us on
ProductSafety@springernature.com

In case Publisher is established outside the EU,
the EU authorized representative is:
Springer Nature Customer Service Center GmbH
Europaplatz 3, 69115 Heidelberg, Germany

Printed by Libri Plureos GmbH
in Hamburg, Germany